Praise for *Shame Nation*

"Engaging, sharp, and important—*Shame Nation* will inspire you to open your eyes and be better in society's growing cyber culture."
—Theresa Payton, CEO of Fortalice Solutions and deputy director of intelligence on *Hunted*

"We all have the power to create the inclusive Internet reality that we so desperately need. Scheff recognizes that participating in that creation can be overwhelming and intimidating, especially given the personal risks we face as women who want to join a public discourse, and provides in *Shame Nation* a clear-eyed, approachable guide to facing a hostile online environment while maintaining our dignity and sanity."
—Emily Lindin, founder and director of the UnSlut Project

"*Shame Nation* is a book that illustrates the harm our culture of shame can wreak when we fail to instill a sense of empathy and kindness. Sue Scheff's advice, couched in examples from current events, has already changed the way I approach social media. Her words have caused me to stop, think, and approach my online relationships from a place of respect, even when I'm angry or apt to reach for the keyboard in retribution for a perceived slight. *Shame Nation* is a practical and helpful book that forges a way forward through the gauntlet of social media and online shaming while encouraging an ethical approach to an increasingly judgmental world."
—Jessica Lahey, *New York Times* bestselling author of *The Gift of Failure*

"With online cruelty and shaming at an all-time high, the Internet can be hurtful and harmful, leaving irreversible damage on its victims and their families. But rather than shrugging our shoulders and closing our eyes, we can empower ourselves through the practical advice and protective measures outlined in *Shame Nation*. Not only does Sue Scheff show us ways to respond when digital damage occurs, she can help us prevent it. I will be recommending this book to everyone I know. *Shame Nation* holds that elusive key to stopping the trend of online hate so kindness and compassion can prevail."

—Rachel Macy Stafford, *New York Times* bestselling author of
Hands Free Mama, Hands Free Life, and *Only Love Today*

"Relatable, intelligent, and engaging from the first sentence, Shame Nation sheds much-needed light on our current culture of online shaming and cyberbullying. Thoroughly researched and packed with eye-opening anecdotes, *Shame Nation* will help you learn why people choose to shame one another online, and what to do if it happens to you or a loved one. Sue Scheff is an invaluable resource in the digital world, and this book should be required reading in high school, college, and the workplace."

—Katie Hurley, LCSW and author of *No More*
Mean Girls and *The Happy Kid Handbook*

"Smart. Timely. Essential. *Shame Nation* is the era's must-read to renew Internet civility where humiliating, criticizing, and judging are normalized. And there is no one better than Sue Scheff to share how to prevent as well as rebound from digital shaming. Her own battle against cybershaming riveted the world and changed the way we use the Internet. Scheff's sage advice and realistic tips for cyber etiquette are indispensable and must be heeded for any hope of a civil existence and restoring empathy in cyberspace."

—Michele Borba, EdD, educational
psychologist and author of *UnSelfie*

"This book offers readers an important and eye-opening exploration into how public shaming transforms lives in the digital age. By humanizing today's digital interactions, Scheff and Schorr show readers hot to recognize, protect, and, when needed, restore their digital identity."

—Stacey Steinberg, legal skills professor at
University of Florida Levin College of Law

"Hardly a day goes by without another vivid example of online hate or cyberbullying in the news. Sue Scheff is determined to put an end to this culture of online shaming with her new book. Brimming with captivating stories of online abuse that keep you hooked, the real rewards of this book are its powerful tips aimed at keeping you from becoming a victim and, more importantly, sage advice on what to do should you find yourself the target of any online abuse. This book is the must-read survival guide for anyone who uses the Internet today."

—Diana Graber, founder of CyberCivics
and cofounder of Cyberwise

"There is no better time than the present to deal with the issues that Sue Scheff presents in *Shame Nation*. Finally there will be sage advice and necessary tools to understand and help each other live in the society-shaming environment we all share."

—Stacey Honowitz, author, CNN legal
analyst, and sex crimes prosecutor

"To create a kinder and more tolerant society, each of us must do our part. This incredible work by Sue Scheff illustrates why—and more importantly—how! *Shame Nation* isn't just a must-read for all parents and teachers, but for every human who desires a more thoughtful and empathetic community in our global village."

—Richard Guerry, author and executive director of the Institute for Responsible Online and Cell Phone Communication

"A leading expert in the digital world, Scheff offers the latest insight as to why people publicly shame each other and will equip readers with the tools to protect themselves from what has now become the new Scarlet Letter."

—Ross Ellis, founder and CEO of STOMP Out Bullying

"Filled with compelling and relatable stories, interviews and quotes from celebrities and experts, and practical strategies, expert Sue Scheff's new book powerfully portrays the all-too-real phenomenon of online shame. Either it has happened to us or to someone we care about—and so much of what is conveyed connects immediately and viscerally with the reader. But there are solutions, and there is hope—as she explains. Take the time to read it."

—Dr. Sameer Hinduja, codirector of the Cyberbullying Research Center

SHAME N▲TION

The Global Epidemic of Online Hate

SUE SCHEFF

with MELISSA SCHORR

Foreword by MONICA LEWINSKY

Published by Sourcebooks, Inc.
P.O. Box 4410, Naperville, Illinois 60567-4410
(630) 961-3900
Fax: (630) 961-2168
www.sourcebooks.com

Library of Congress Cataloging-in-Publication Data

Names: Scheff, Sue, author. | Schorr, Melissa, author.
Title: Shame nation : the global epidemic of online hate / Sue Scheff, with Melissa Schorr.
Description: Naperville, Illinois : Sourcebooks, [2017] | Includes bibliographical references and index.
Identifiers: LCCN 2017013493 | (hardcover : alk. paper)
Subjects: LCSH: Cyberbullying. | Shame--Social aspects. | Humiliation. | Online hate speech. | Internet--Moral and ethical aspects.
Classification: LCC HV6773.15.C92 S355 2017 | DDC 302.34/302854678--dc23 LC record available at https://lccn.loc.gov/2017013493

Printed and bound in the United States of America.
MA 10 9 8 7 6 5 4 3 2 1

To my children, Ashlyn and Scott, my grandchildren, Julia and Jordan, and my parents, Robert and Maureen.

CONTENTS

FOREWORD

BY MONICA LEWINSKY

If you picked up this book, you are already aware of the sweeping digital landscape of the Internet littered with harassment and the trolls who infuse online discourse with aggressive shaming and bullying. Targets can be as small as personal attacks on Facebook pages to the explosion of political trolls' tweets who seemed to highjack our political process in the 2016 election.

In *Shame Nation*, author Sue Scheff addresses this onslaught of public shaming by assessing the problem, sharing her own experiences with online shaming and advocating for a better, safer Internet. She recommends available resources for assistance as well as very helpful suggestions for overcoming the pain and humiliation of cybershaming. And Scheff details the heartening changes in the legal sphere where online harassment and cyber-stalking are finally beginning to develop remedies for victims and consequences for perpetrators.

Shame Nation presents a road map of harassment on the Internet: why people harass others online, how they do it, and the various ways one can respond—or vigilantly avoid being the target of online shaming. Essentially, how to be prepared for a virtual attack on one's character.

Read that again: **how to be prepared for a virtual attack on one's character.** That sentiment is heartbreaking. How have we devolved as a society to the point where we even need to be preparing for such an attack? What is it about online communication that propagates such venal commentary? Of course, at its core, power. Power over one's own feelings of inferiority, power over another (perceived) weaker person, power to be seen and heard. Then there is also the mob mentality. Added to that, the safety of anonymity the Internet provides.

And then there is the phenomenon Scheff unpacks later in this book—what psychologist John Suler describes as the Online Disinhibition Effect, wherein online behavior distances us from our normal personalities and encourages us to develop different personas. One only has to observe the myriad of online usernames that range from the fanciful to the outright frightening to know this is true.

When I first became the subject of worldwide shaming in 1998, few people could imagine how that would feel—or what the long shadow of consequences could look like. (Oh, how I wish Sue's book existed then!) Back then, literally overnight, I went from a completely private person whose world was just family, friends and coworkers to a globally known figure, relentlessly mocked and vilified. I stepped (or was pushed, really) from a comfortable, anonymous existence into a maelstrom of wild rumors and cruel jokes. And the nascent Internet played an important role in that storm. In fact, the report that launched this story first broke online in one of the very first online news blogs. All of this was new territory, and there were no rules and no boundaries.

As an advocate who has traveled all over the world speaking about this issue, I have had the vantage point of hearing from countless people about their struggles on this topic. Today, sadly, many people can identify with the author's and my experiences of being shamed by millions of people—millions of strangers. From Sue herself, whose business and reputation was viciously attacked in an online smear campaign to a Canadian high school where a poll called "The Ugliest Girls in the Twelfth Grade" was published to the 38 percent of adults who confirm they have been cyberbullied in the last year. And those numbers are on the rise.

And tragically we have come to a point where we had to coin a new word—bullycide—to describe the onslaught of suicides, especially among young people and children, resulting from bullying, particularly cyberbullying.

In 2014, after a decade long self-imposed retreat from public life, I authored an essay for *Vanity Fair* titled "Shame and Survival." In it I reflected:

> No one, it seems, can escape the unforgiving gaze
> of the Internet, where gossip, half truths, and lies

take root and fester. We have created, to borrow a term from historian Nicolaus Mills, a "culture of humiliation" that not only encourages and revels in Schadenfreude but also rewards those who humiliate others, from the ranks of the paparazzi to the gossip bloggers, the late-night comedians, and the web "entrepreneurs" who profit from clandestine videos.

Yes, we're all connected now. We can tweet a revolution in the streets or chronicle achievements large and small. But we're also caught in a feedback loop of defame and shame, one in which we have become both perps and victims. We may not have become a crueler society—although it sure feels as if we have—but the Internet has seismically shifted the tone of our interactions. The ease, the speed, and the distance that our electronic devices afford us can also make us colder, more glib, and less concerned about the consequences of our pranks and prejudice. Having lived humiliation in the most intimate possible way, I marvel at how willingly we have all signed on to this new way of being.

There is painfully a sad lack of empathy and compassion in our cyberworld. People rush to make rude and (sometimes) violent commentary they would never utter in a face-to-face situation. And these insults and threats never go away. They live in the Internet ether forever, easily accessed by potential employers, potential relationships, and anyone in the mood to do a Google search. But society is slowly acclimating to this "brave new world," and we are gradually developing new methods for coping with online harassment and curbing cyberbullying and shaming in all forms. These include legal recourse against the perpetrators and well as counseling and support for those victimized by online harassment.

Activist and support groups have sprung up globally. The concept of digital resilience has been appropriated in this space to look at ways our social behavior and emotional responses can be reshaped so that we—especially young people—are able to bounce back when (cyber)bullying and online harassment occur. But ultimately, as a society we can—and must—do better to help protect others online, be mindful of our own clicking behavior, and remember neutral compassion. As an advocate in the antibullying space, I think I summed up my thoughts best in an interview with Guy Raz for NPR's TED Radio Hour on why I ultimately have so much faith in our ability to make progress:

> The most important thing that I've learned is that we are all so much stronger and so much more resilient than we can ever imagine. I think it's so important for people to understand that. And it doesn't matter your experiences of struggle or

level of trauma, we are more resilient, we have
a well of compassion for others and ourselves,
which can help us bounce back from any kind
of situation.

The reason Sue called this book *Shame Nation* is because
shame culture doesn't make value judgments on one's actions,
but instead, more insidiously, it tells people that they, as human
beings, are unworthy. This most valuable book is an attempt to
frame the human stories on either side of (cyber)bullying—to
strip away the screens and digital posturing and create a narrative
steeped in empathy. Let's begin...

—*Monica Lewinsky*
New York City, 2017

INTRODUCTION

THE CLICK: WHAT IS DIGITAL SHAMING?

"There is a very personal price to public humiliation, and the growth of the Internet has jacked up that price."

—Monica Lewinsky[1]

Do you know, right now, what the Internet is saying about you?

Could one careless tweet cost you your job? Are nude photos of you lingering on your ex's smartphone? Could one angry customer trash your small business? Will a potential romance cool because of what's been posted about you online?

How likely is it that any of that will happen?

More likely than you think.

In today's digitally driven world, countless people are being electronically embarrassed every day. Stories of troll attacks, revenge porn, sexting scandals, email hacks, webcam hijackings, cyberbullying, and screenshots gone viral fill our newsfeeds. According to a 2014 Pew Research Center survey, 73 percent of adult Internet users say they have witnessed online harassment and 65 percent of adult Internet users under the age of thirty have personally experienced it.[2] Given events like the 2014 Sony Pictures email hack that leaked studio heads' private messages and the 2015 Ashley Madison breach that revealed the identities of millions of alleged philanderers, it is clear that we are all potentially one click away from being unwillingly thrust into the Internet glare.

And what awaits us there? A nation of finger-wagging vultures who delight in tormenting us and tearing our reputations to shreds. This culture of destroying people with the simple stroke of a keyboard has become much more than a fad—it's the new normal. In a 2014 survey conducted by YouGov, 28 percent of Americans admitted to engaging in malicious online activity directed at somebody they didn't even know.[3] How have we become this "Shame Nation," where we are constantly hurling our collective outrage at an endless supply of fresh victims? And is there anything we can do to stop this, before it affects us or the people we love?

Of course, shaming in America dates as far back as the days of the Puritans, when those deemed to have crossed their thin moral line were subject to being stoned, scorned, thrown into stocks, or worse. Just a generation ago, an embarrassing gaffe might have been written up in the local paper or gossiped about over backyard fences until it was old news. But today is much different. The Internet has eternal life and boundless reach, and victims of a digital disaster must learn to live forever with the implications of that high-tech "tattoo." As Jennifer Jacquet, an assistant professor of environmental studies at New York University, writes in her book *Is Shame Necessary?: New Uses for an Old Tool,* "The speed at which information can travel, the frequency of anonymous shaming, the size of the audience it can reach, and the permanence of the information separate digital shaming from shaming of the past."[4]

In other words, being the victim of a cyberattack has the potential to ruin your life—financially, emotionally, and physically. In the most extreme cases of online harassment, we have seen the worst-case repercussions over and over again: young people taking their own lives and adults losing their livelihoods.

Strikingly, Chubb Limited recently became the first insurance firm to offer its clients coverage against cyberbullying and other forms of man-made digital disasters. For $70 a year, American families can add a protection plan to their existing policy and get reimbursed for up to $60,000 in costs resulting from online harassment, such as unwarranted job loss, technical support for tracking down cyberfoes, public relations support for image repair, and even therapy bills.[5] "This is a low-probability but high-consequence issue," explains Christie Alderman, Chubb's vice president of client product and services. "It's not something that's happening every day. But when it does, it has a huge impact on people."[6]

Take the case of the chief financial officer of an Arizona-based medical device company who, in 2012, decided to take a stand against fast-food chain Chick-fil-A, which at the time held a controversial stance against gay marriage. He filmed himself berating a drive-through worker and shared the video on his YouTube channel. The clip went viral, costing him his job—and, he claims, leaving him unemployable *for years*. In a follow-up story by ABC News in 2015, he told the reporter he was on food stamps, had been living with his family in an RV, and was still looking for full-time work.[7]

Would you be prepared if such a virtual catastrophe were to happen to you?

From Victim to Expert

In 2003, vindictive cyberattackers crashed into my world and tried to destroy my reputation, my livelihood, and my serenity. After a spiteful client decided to go after me and the Florida-based organization I founded to advise parents of troubled teens, Parents' Universal Resource Experts (P.U.R.E.), I found myself on the receiving end of a brutal smear campaign. Armed with a keyboard, anonymity, and the belief that the First Amendment grants carte blanche to vilify others, the campaign cascaded into a free-for-all—with me up for slaughter. I was called a crook, a con artist, and, worst of all, a child abuser. I was a forty-year-old single mom left on the verge of financial ruin, feeling completely desperate and alone. With those slurs attached to my name, I felt certain that no one—clients or even potential dates—would ever take a chance on me again. I became a total recluse, living in fear of people Googling me.

As I detailed in my previous book, *Google Bomb*, instead of shrinking from the world, I fought back. I sued for Internet defamation and invasion of privacy and won a landmark, $11.3-million-dollar judgment that made national, and even international, headlines. Unfortunately, although the courts cleared my name, the Internet (almost) never forgets. I have spent years rebuilding and maintaining my online reputation, but even today, there is still slime lingering about me in cyberspace.

As horrible as the experience has been for me personally, it did have one positive outcome: I have become one of the nation's leading advocates for responsible digital citizenship and cyber-safety. Major media outlets, such as CNN, the *Washington Post*, Fox News, *20/20*, the *Wall Street Journal*, and others, call me when news hits of another major cyberlynching. In addition to the tools and strategies I learned during my own case, I have sought out top experts in such fields as law, technology, and psychology. Using all that I've gleaned, I've made it my mission to help others create a safer and healthier digital life. As an adult victim and survivor of cyberbullying and digital shaming, I personally understand the deep, dark hole into which you descend when you are attacked. My greatest satisfaction now comes from helping others reclaim their lives after they, too, have been cyberbombed.

In these pages, I attempt to survey the wreckage of the most shocking digital debacles in recent years, revealing how truly pervasive this phenomenon has become and advising what we as concerned citizens need to do about it. (Because this book is about education rather than exploitation, names have been changed or removed as appropriate.) You will find expert advice about how to avoid making similar mistakes so you don't become a victim yourself, plus stories of inspiration from those who have survived. Finally, this book will provide practical solutions for how to handle a cyberdisaster, should one occur, and guidance on how to recover from the emotional aftermath of digital shaming. You can find new life after falling victim to today's legal—yet, lethal—weapon: the keyboard. Following is a brief overview of some of the major concepts and stories that we'll examine in this book.

The Rise of Shame Nation

Who are today's victims of digital shaming? Potentially, any one of us. Anger your ex, and your nude photos could show up on MyEx.com, a website dedicated to the practice of revenge porn. Bad tipper? You could be dubbed a cheapskate on your local deliveryman's Tumblr page.[8] Cut off a fellow mom at the drop-off line at school, and you could find yourself trashed on Facebook later that morning.

Danielle Keats Citron, author of *Hate Crimes in Cyberspace*, estimates that 30 to 40 percent of us will experience digital shame at some point in our lifetime. "You never escape it," Citron, a law professor at the University of Maryland, told the *New York Times*. "When you post something really damaging, reputationally damaging, about someone online, it's searchable and seeable. And you can't erase it."[9]

And don't think you can somehow keep yourself from being victimized by keeping a low profile. Countless bystanders have been thrust into the Internet's hot public glare without even knowing it. Imagine going to the hospital for urgent medical care and finding out that your nurse had snapped and texted photos of your private parts to her colleagues for a laugh.[10] Or that paramedics whisking you away in an ambulance were engaged in a cruel competition, taking incriminating shots while you lay unconscious.[11]

More famously, there is the case of #PlaneBreakup, where one rather unfortunate woman's boyfriend tried to dump her while they were stuck together on an airplane. By the time they landed, thousands of people had virtually eavesdropped on their semiprivate conversation, and a photo of her sobbing had been snapped and uploaded to the world—receiving ten thousand likes. All because a fellow passenger decided to live tweet this lovers' quarrel for the Twitterverse's amusement.

I Can't Believe They Posted That!

Of course, many of today's digital debacles are of our own making. There are countless high-profile examples of people whose own words unwittingly brought disaster tumbling down upon their heads. One of the best-known cases involved Justine Sacco, a public relations director from New York who was flying to South Africa and whose sardonic tweet—"Going to Africa. Hope I don't get AIDS. Just kidding. I'm white!"—set off one of the first ever Twitter lynchings. Since then, there has been a flood of these stories, each predictably triggering a massive cybershaming in response: the high school teacher from California who joked about wanting to stab her students, the New Year's Eve restaurant patron whose Facebook rant backfired on her, the Harvard professor's email feud over his Chinese take-out order.

Just as we can bring ourselves down with our own thoughtless or misconstrued words, we also take risks every day with the images that we share. Consider the fact that every **second**, more than

+ 3,472 pictures are uploaded to Facebook,
+ 66,000 people search using Google,
+ 5,700 tweets are posted,
+ 926 pictures are uploaded to Instagram,
+ 8,796 "snaps" are sent, and
+ 400 minutes of video are uploaded to YouTube.[12]

When it comes to sending photos with nude or risqué content, the stakes have never been higher. According to a 2015 survey conducted by the Canadian research firm Leger, one in five parents in Canada admit to sharing intimate photos and/or messages online or via text—despite today's high rates of divorce.[13] And yet we expect our hormonally charged, tech-addicted youths to restrain themselves? Statistics show that 20 percent of teens and 33 percent of young adults have posted or sent nude or seminude photos, which can be a pathway to an embarrassing ending—witness the sexting scandals that have popped up in small towns across America, from Duxbury, Massachusetts, to Cañon City, Colorado.[14]

Adults can get tripped up too. Just ask Golden State Warriors basketball star and Team USA member Draymond Green, who made a mistake of Olympic proportions in July 2016, when he (briefly) posted a picture of his private parts on his Snapchat Story for all the world to see. "I kinda hit the wrong button and it sucks," he was forced to tell reporters and the world in a Team USA press conference later that day. "It was meant to be private. We're all one click away from placing something in the wrong place—and I suffered from that this morning."[15]

A Sadist's Playground?

"The Internet is the ugliest reflection of mankind there is."

—Iggy Azalea[16]

Kids telling other kids to "drink bleach and die." A troll creating a fake Twitter account posing as a woman's dead father. Mothers being told that their babies should die of sudden infant death syndrome (SIDS).[17] Truly, the breadth of human cruelty on digital display can be staggering. As the former CEO of Reddit once remarked, reading online insults made her "doubt humanity."[18]

As we all witnessed throughout the 2016 U.S. presidential campaigns, the level of civil discourse in our country has never sunk quite so low. "We are in an age of incivility," observes longtime tech journalist Larry Magid. "I think that's pretty obvious. The trolling and harassment and disagreeable conduct we've seen over the last several years has contributed to the ugliness, not so much its influence on a candidate, but on the broader community. There is a vocal group of people who seem to have forgotten their manners."[19]

Why *are* people so mean online?

And what are the chances that you will personally brush up against such a foe?

Unfortunately, there does seem to be a small subgroup of citizens who simply enjoy engaging in digital combat. According to the 2014 study "Trolls Just Want to Have Fun," published in *Personality and Individual Differences*, 5.6 percent of those surveyed admitted that their favorite activity when posting online comments is trolling others—and not surprisingly, they also scored high on traits such as sadistic and psychopathic tendencies.[20] But Lindsay Blackwell, a University of Michigan researcher who studies online harassment, says that we should not rush to call those partaking in online cruelty psychological outliers. "I think it's dangerous to flip that around and say everyone who participates in these behaviors has one of those traits," she said in an interview. "We're all equally capable."[21] In other words, push the right button, and there's a potential troll lurking inside us all. This was validated when Stanford and Cornell University released research in 2017 confirming that under the right circumstances, *anyone can become a troll.*[22]

"We have to understand the human condition," explained nationally renowned psychologist Dr. Robi Ludwig, author of *Your Best Age Is Now*. "Human beings have that aggressive, murderous side to them, and it's not going away. The human species is aggressive, and it will manifest itself in different ways. Right now, it is taking this form, with people sitting alone at their computer feeling frustrated or insecure. How easy is it to turn that aggression to attacking words?"[23]

Psychologists point to several factors that have allowed online cruelty to flourish:

+ the anonymity of the Internet
+ the distance, or lack of face-to-face contact, with a victim
+ mob mentality run amok
+ lack of gatekeepers
+ lack of consequences

Taken altogether, this phenomenon is known as the online disinhibition effect, the notion that people behave far differently online than they would in reality. John Suler, PhD, a psychology professor at Rider University who was the first to formally tackle this issue in a 2004 research paper, explains that the lack of a physical link between the attacker and the victim makes it easier to say things one wouldn't in person. When this disinhibition turns toxic, or toward attacking others, it could be for several reasons: the online poster may not know exactly who the victims are or see them as real people, there are few consequences for this nasty behavior, and there is always the ability to hop off the discussion at any time. "We live in an age when people feel frustrated and angry," Suler observed in an interview. "The online disinhibition effect causes people to act out that frustration and anger."[24]

"The Internet itself is partly to blame," Ari Ezra Waldman, associate professor and director of the Innovation Center for Law and Technology at New York Law School, writes in a *New York Times* editorial on the problem of online harassment. "Much of what makes it great—its speed, low cost, scope, size, and pseudonymity—also facilitates avalanches of hate."[25]

It's clear that hiding behind a computer screen enables many to express themselves in ways they'd never dare to face to face. But are online trolls becoming so brazen that they're beginning to shed even those inhibitions? Perhaps. One ominous finding, in a 2016 study by researchers at the University of Zurich, reported that people making hateful comments on an online petition website were likely to behave even *more* aggressively if they identified themselves with their full names than if they remained anonymous.[26]

Why are trolls becoming more willing to publicly take an unpopular stance? The researchers suggested that these posters may have found that they could win more like-minded followers if they were willing to stand behind their words. And sadly, these haters are finding that most—though, as we will see in the pages ahead, not all—of the time, they are immune to any real-world consequences.

Signing Off Is Not a Solution

So is all this to say that you shouldn't have any online presence at all? Should you shut down your Twitter account, log off Facebook, and delete your LinkedIn profile? Wipe your Instagram photos, turn off your wireless, and, for good measure, hide out in your bedroom? Of course not. If you don't maintain your digital profile, who will? Going off the grid is just not an option—and it won't save you. Those who have tried opting out of social media as an experiment often report feeling disconnected and even ostracized from their social circles. As journalist Nancy Jo Sales, author of the book *American Girls: Social Media and the Secret Lives of Teenagers*, relates, "I spoke to girls who said, 'Social media is destroying our lives. But we can't go off it, because then we'd have no life.'"[27]

From a professional standpoint, if you lose your job before you're financially ready to retire, or if you're simply looking for something new, you'll be light-years behind other job-seekers if you haven't maintained your virtual footprint. Having no online history is just as risky as having a spotty one—employers wonder what you have to hide or assume you aren't tech savvy.[28] A 2016 CareerBuilder survey found that 41 percent of employers were less likely to interview candidates if they couldn't find them online.[29] Many careers now require us to participate publicly in social media and to maintain a digital presence. As Congresswoman Katherine Clark, one of the leading proponents for federal laws against cyberharassment, expressed in an interview, "In an economy that demands an online presence, we can no longer view online abuse as simply a virtual problem."[30]

For better or worse, being a vocal member of our digital nation is practically a modern-day requirement. That's why more than ever, we need to learn and begin practicing what I call digital wisdom, to prevent ourselves and our loved ones from being victims of a digital debacle—and to make sure that what we are putting out there reflects the online world we all desire to live in.

Can we turn our Shame Nation into a Sane Nation?

Let's begin.

PART ONE

THE RISE OF SHAME NATION

CHAPTER 1

THE SPECTRUM OF SHAME

"So many people want their fifteen minutes of fame— and don't care how they get it."

—Richard Guerry, founder of the Institute for
Responsible Online and Cell Phone Communication[1]

The Sport of Shaming

Imagine this scenario: you're on an airplane, heading home after a long weekend away, when the pilot announces that your flight has been delayed. You're stuck on the tarmac, bored and annoyed, when a couple across the aisle begins arguing.

The fight gets heated. Tempers flare. Tears are shed. Accusations are hurled.

What do you do?

Most of us would just put on our earbuds and bury our heads in a magazine.

One woman made a different choice. She pulled out her phone and started posting the couple's drama, verbatim, to the world.

Welcome to #PlaneBreakup.

That was the Twitter hashtag that trended in August 2015 when Kelly Keegan, a twenty-four-year-old ad sales associate from New York City, began live tweeting every word of her fellow passengers' fight—along with their subsequent makeup make-out session—during a flight delay on a plane headed from Raleigh, North Carolina, to Washington, DC. Here's just a snippet of what she wrote:

This guy on the plane just broke up w his girlfriend and she's SOBBING

↰   ⇄   ♥   •••

Girl: "ITS JUST SO MEAN. DO I DESERVE THIS? WHY ARE YOU BRINGING THIS UP"

↰   ⇄   ♥   •••

Guy: "You need to calm down"

↰   ⇄   ♥   •••

Girl: "To me I just really thought, you know, this was going to go somewhere"

↰   ⇄   ♥   •••

And then, Keegan's assessment of the in-flight experience.

What made Keegan do it? What drives our desire to shame others so harshly—even strangers? And what are the potential costs when we do?

Keegan, a Jersey-born, former Kappa Delta sorority girl at North Carolina's High Point University, agreed to be interviewed and shared what her thought process was that fateful Sunday evening. As the argument grew in volume, everyone on the plane began paying attention. Keegan, who was flying back from a girlfriend's out-of-town bachelorette party, recalls the pivotal moment when she inserted herself into the situation. "I had cell service, so I said, 'why not?'"

Keegan had been on Twitter since 2009, largely for the celebrity gossip, but at the time had only had around six hundred followers. Her tweets were amplified when they were picked up by friends working at the popular website, Barstool Sports. "They started getting involved and laughing and retweeting me, then people in media started doing the same," she says. By the time she turned off her phone as the plane was taking off, she had around seven thousand followers, and by the time she landed, she was in the midst of a media frenzy. When she reached her apartment at midnight, her followers had grown to more than twelve thousand, and she was fielding calls from the *New York Post* and BuzzFeed.

Not everyone was delighted. "My mother," Keegan observes wryly, "was horrified." But Keegan and those who joined in on the mockery felt they had some justification for this in-flight shaming. The couple had violated certain basic norms of public behavior—don't air your dirty laundry in public, don't get drunk and disorderly, and don't aggravate your fellow airline passengers, who are already suffering cramped leg room and recycled air.

From Keegan's perspective, social outliers who behave this way do so at their own risk. "People need to be more self-aware [about] what they're doing and how they're acting in public," she says. "It wouldn't happen to me, because I don't act that way. I'm never one to create a scene—that's how I was raised. People act as if [they] can spew nonsense into this void and have zero repercussions. I've obviously proved that's not true. That's what happens—someone like me comes along who is nosy. Why would you not assume, all the time, [that] there's a possibility someone's looking at you? Everyone is watching. If you're bored on a plane and arguing right next to me, what am I supposed to do?"

Clearly, Keegan is not the only one irked by airline passengers behaving badly. Witness the wildly popular "Passenger

Shaming" Facebook and Instagram accounts, where a former flight attendant posts images of air travelers' transgressions— painting their toenails or dozing bare-chested—submitted by their frustrated seatmates.[2]

Yet at the same time, Keegan herself had violated other, equally valid, community norms—don't stick your nose in someone else's business, don't ridicule strangers for sport, and, some argue, show some basic compassion for people who are clearly in crisis mode. Within days, Keegan's behavior had generated its own backlash: online commenters who'd read the story on People.com and USweekly.com began calling her callous and evil. "I started seeing people commenting [on] how I should have been minding my own business," she recalls. "The worst was, 'I hope someone live tweets you crying at your grandma's funeral.'" A friend tipped her off that someone had even created a fake Facebook profile of her, mock apologizing. The pile-on toward her also took a real-world turn. Keegan's boss, at the job she held before taking her current position, began getting calls asking him how he could employ such a "despicable" person.

Most of this blowback didn't faze her. "I'm lucky—I'm a person who ignores people shouting into the wind," she says. "I've never taken these super to heart." One year after #PlaneBreakup, she has retained some thirty thousand Twitter followers and hosts a popular weekly podcast called *WhineWithKelly*. The couple from the plane has never come forward or contacted her, and she insists that it was just a case of "mouthing off on the Internet." "People make the argument, 'You hurt these people,'" she says. "I think if I had really impacted their lives in any way, they would have reached out to me and called me an asshole."

When asked if she would do it all over again, she briefly pauses to think. "Um… I wouldn't say no. I personally like having this audience, I like being able to provide content for them and try[ing] to make it funny." She does have one partial regret: "I don't think I would include photos next time. That was a lot of backlash. None of it was illegal, but [it] definitely had me worried for a little while."

But otherwise? "I think if an opportunity arose," Keegan says, "I would react the same way."

Shame, Defined

1. a painful emotion caused by consciousness of guilt, short-coming, or impropriety
2. a condition of humiliating disgrace or disrepute
3. something that brings censure or reproach[3]

How have we reached this state of being a shame nation, where nothing and no one is off limits? From celebrity babies to seniors in nursing homes, no one seems immune from becoming a victim of a virtual whipping—encounter the wrong person and you, too, can become a target.

Shaming is now the weapon of choice everywhere we look. We see it in arrogant bodybuilders mocking out-of-shape gym members.[4] Middle school students creating fake Facebook pages to trash their teacher or principal. Thousands of "tough-love" moms and dads filming YouTube videos harshly scolding—or even shearing their children's hair on camera in a last-ditch attempt at discipline. Websites like Yelp and Pissed Consumer that allow angry patrons to publicly blast a small business owner, fairly or not. A never-ending churn of popular apps, from After School to Yik Yak to ASKfm to Burnbook, permit teens to slam one another anonymously, while the controversial app Peeple encourages adults to "review" others.

The list goes on and on.

Whether the motive is profit, revenge, or simply sport, so many of us feel entitled to pick apart every aspect of a person and disparage it. We judge their way of life, their parenting style, their religion or sexual orientation, how they look, and how they dress. And all too often, the entire community either tacitly approves or gleefully joins in. To be truly shamed, you need an audience witnessing the deed. *New York Times* columnist David Brooks writes about the distinction between shame culture and guilt culture: "In a guilt culture you know you are good or bad by what your conscience feels. In a shame culture you know you are good or bad by what your community says about you, by whether it honors or excludes you. In a guilt culture people sometimes feel they do bad things; in a shame culture social exclusion makes people feel they *are* bad."[5]

We shame to pressure outliers to conform to our norms— even if no one can agree anymore what those standards should be. "I think a lot of people resort to public shaming out of anger and frustration, the desire to call out bad behavior, and the need to feel validated for their emotions," writes Christine Organ in an essay on shaming published on the blog *Scary Mommy*. "We feel justified in sharing that photo or video, entitled to call out the rude, crass, or inappropriate behavior... We're doing the world a favor, thankyouverymuch."[6]

"Shaming is not new," observes *Today* show contributor and educational psychologist Michele Borba, EdD. "But what's different is that it's slowly become almost part of acceptability—the new norm. It goes with the same issue of [the] breakdown of civility [and] respect, [the] diminishment of empathy—the perpetrator doesn't see anything wrong with it ('Everyone else does it; she deserved it; what's the big deal?') because he's not considering, 'How would I feel if that happened to me?'"[7]

Ironically, sometimes shaming others can make the shamer look good. Research suggests that some people find a specific benefit in shaming—calling out bad behavior in others makes you appear more virtuous. Researchers at Yale University wanted to find out why humans care about selfish behaviors that do not affect them personally, using game theory to explore this particular puzzle, known as third-party punishment. In their 2016 study, published in *Nature*, they concluded that shaming others can boost our own reputation and signal to the larger group that we are not selfish ourselves. "Sometimes," they wrote, "punishing wrongdoers is the best way to show that you care."[8]

Shaming for Good?

Indeed, shame sometimes can be used as a force for a greater public good: say, when critics bring attention to alleged price gouging in the pharmaceutical industry, as was the case with Mylan's EpiPen, or when police departments' Facebook pages publicly name individuals caught drug dealing or committing sex crimes. Irate consumers, frustrated by shoddy customer service, have effectively turned to social media to air their grievances, like when Megyn Kelly turned to Twitter to blast Shutterfly about her Christmas card order, claiming the company lied about her order for weeks.[9] Even in modern times, judges occasionally revive centuries-old shaming tactics, ordering petty criminals to parade around with a sign around their necks describing their misdeeds. "Shame's service is to the group, and when it is used well and at the right time, it can make society better off," Jennifer Jacquet writes in *Is Shame Necessary?*. "When shame works without destroying anyone's life, when it leads to reform and reintegration…or, even better, when it acts as a deterrent against bad behavior, shaming is performing optimally."[10]

Even the U.S. federal government has dipped a toe into the business of shaming its citizenry. In April 2016, the National Highway Traffic Safety Administration created a #JustDrive Twitter campaign against distracted driving, scolding—by name—individuals who joked or even bragged about texting and driving. For a period of time, the agency's social media guru would look for mentions of texting and driving and then insert himself into the conversation, aggressively calling out those scofflaws.

When some lady honks at me for texting and driving and I flip her off without looking up from my phone 😎 #multitasking

NHTSA @NHTSAgov

If you're think you're being cool, @ ▓▓▓▓▓▓▓, you're not. Put the phone down and #justdrive, please. It's not worth it.

I'm either texting and driving, eating and driving or sometimes both at the same time 😂

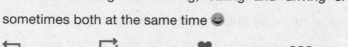

NHTSA @NHTSAgov

Not funny, @ ▓▓▓▓▓. Put down the phone and #justdrive, please—the world's a better and safer place when you do.

Is this an appropriate use of resources by our federal government? According to the Department of Transportation, the edgy campaign elicited $8.9 million in donated media coverage.[11] "[It] could be referred to as 'shaming,'" says Lori Gabrielle Millen, marketing specialist for NHTSA, "but we like to think of [it] as 'educational messaging.'"[12]

In some situations, shaming is all we have.

The name Brock Turner first entered the national conscious-ness when the Stanford undergraduate was convicted of raping an unconscious twenty-two-year-old woman behind a frat-house dumpster. The anonymous victim wrote an impassioned letter that managed to stoke an entire nation's sympathy. Outrage only grew after Judge Aaron Persky doled out a paltry six-month sentence (Turner ended up serving only three months). "Shame on you," one irate juror supposedly wrote in a letter to the judge.[13]

Stymied, we the people felt that our legal system had failed. We raged, we blogged, we vented, but finally, all that was left was to enact vigilante justice. And so Brock Turner was essen-tially sentenced to social media jail, ensuring that at least his reputation would be destroyed for all eternity. Upon his release three months later, the shaming resumed, with Facebook memes of Brock's smiling head shot, superimposed with the words MY NAME IS BROCK I'M A RAPIST, reaching more than a million shares.[14] Facebook pulled some of the posts, citing its antibul-lying policy, then reinstated them, stating that protection does not apply once people have entered the "public interest."

Shaming was all the public had, and it will have to be enough.

Shades of Shame

But the shades of shaming are not always so clear-cut. What should we make of the thousands of parents who have taken to social media, posting videos where they purposefully shame their own children? Can shaming have pros and cons all at the same time? The practice is worrying to parenting experts who point to the potential psychological damage these children may suffer. As Lisa Damour writes in her book, *Untangled: Guiding Teenage Girls Through the Seven Transitions into Adulthood*: "Shame is one of the last places we, as parents, want to land with our kids… Shame has toxic, lasting effects and no real benefits."[15] One of the highest profile cases involved a Denver mom who tried this tactic on her thirteen-year-old daughter. In her five-minute video posted on her daughter's Facebook page, she takes her daughter to task for claiming to be nineteen years old and posting salacious pictures of herself.[16]

On the one hand, it's hard not to admire this mom's determination to protect her child and prevent her from growing up before she's ready. Many, many viewers praised the tough-love mom's use of shaming. On the other hand, the video, which went on to receive 11 million views, is difficult to watch.

In it, the girl, wearing a turquoise T-shirt and long pants, speaks softly, shuffles her feet, tugs on her hair, and looks like she'd give anything to be anywhere else on earth. "You're thirteen! So why does your Facebook page say you're nineteen?" demands her furious mom, who tells viewers that the girl still has a bedtime of ten o'clock, still can't "wipe her own ass," and still wears panties that say the day of the week on them. "Tell them you still watch Disney Channel!" the mother crows as the girl slips into tears, repeating the words as she attempts to hide her face.

In a follow-up video, the single mom thanks those who supported her, saying, "It was really hard for me to do, but I didn't want to be another parent on Facebook putting out a video where I beat my child or anything like that... I wanted it to be something that showed from one mother to another mother, to the fathers out there struggling trying to raise a child or a teenage child, just to get them to be aware and to understand how serious it is and how important it is to be aware of what your child is doing at all times."[17] She also explains that she herself had struggled with the law after a drug charge landed her with a criminal record, and that she wants to make sure her own daughter does not go down that path. "I would rather embarrass her and do this than go to a morgue and verify my child's body," she later told ABC News.[18]

Will this strategy work? The answer is unclear. According to a 2016 article on shaming in *Scientific American*, those who internalize messages like this and truly feel badly for their behavior are more motivated to repent and improve than those who simply worry about how their reputation will fare.[19] It's also hard not to wonder about a potential fallout between the parent and child. As Damour points out, "Once shamed, teens are left two terrible options: a girl can agree with the shaming parent and conclude that she is, indeed, the bad one, or she can keep her self-esteem intact by concluding that the parent is the bad one. Either way, someone loses."[20]

For this child and countless others, only time will tell.

Shaming Run Amok?

How do we know when shaming has gone too far? In May 2016, Etan Thomas, the dreadlocked, six-foot-ten former NBA star of the Washington Wizards, boarded an Amtrak train at New York City's Penn Station and asked a young white female if the empty seat next to her was free. When she replied that it was taken, he sat nearby. Minutes later, a young white man asked her the same question—but this time, she said the seat was available. An irate Thomas confronted the woman on her motivations. "She gave a deep breath, like she was so bothered, all disgusted," he later recalled on his sports radio show, *The Collision*.[21] "I didn't let it go." The exchange between the two grew heated, with her denying that race had played a role, even claiming that she'd had an African American boyfriend in college.

Eventually, Thomas whipped out his cell phone and snapped a photo of her, against her protests. "I said, 'I'm going to take this picture and put it on Facebook, because people don't think this stuff happens,'" he said. "I never imagined my little Facebook post would go viral."

But it did, of course, generating widespread condemnation of this anonymous woman's actions. "Twenty-five people have sent me her name—the power of social media is amazing,"

Thomas said. "Some people loved I said something, some despised it, like 'How *dare* you. This poor innocent white woman that was just violated'... A lot of people are saying, 'Well, I hope she sues you.'"

Who knows the full story here? Was the woman really racist? Did she prefer not to have someone with such a large physical presence manspreading all over her personal space? Did she get a text, and her plans changed? Was she a seat hog, the type who angles to sit alone unless repeatedly pressed? It's impossible to know. So far, Thomas has kept mum as to her identity, but one wonders who has seen the photo—her boss? her family?—and whether there have been any repercussions.

Professor Jacquet warns that we "must be mindful" of digital shaming's "power and its liabilities," *especially* when dealing with individuals. "We've always had gossip as a form of shaming, but it now has such scale and speed and the results can be pretty ugly and undesirable," she told the *Guardian.* "We need to take a step back and ask: 'Is this how I want to spend my attention?'; and, 'Who deserves to be a victim of something this severe?'"[22] When I spoke with her, she explained, "I really have seen cases where people have gotten it wrong. We've destroyed this person's reputation. This is not really serving the broader good at all. I think it is about disproportionate punishment—any time the punishment outweighs the crime."[23]

Some casual observers of the Thomas incident seemed to agree. "Publicly shaming her, especially considering his position as a celebrity/media figure and the ensuing wide net of negative coverage that will harm her personally, and possibly socially and professionally, was several steps too far," wrote one commenter on a *Washington Post* article.[24]

We've entered a gray area where so many of us are quick to blame and slower to consider. Where every move we make when we're out and about in public can be accused, tried, and convicted by an angry jury of one.

Or thirty.

In the summer of 2016, a group of thirty cyclists in Charlotte, North Carolina, were out on their weekly bike ride when they encountered a woman driving a silver Mercedes-Benz E350 sedan, with a young girl, presumably her daughter, sitting beside her. Was this Mercedes Mom in a rush for some reason? Or was she just having a bad day? Irritated by the mass of bikers, she buzzed them closely in the bike lane, then jammed on her brakes directly in front of them, putting them at risk of crashing into her. When the cyclists caught up with her at a red light and confronted her about her road rage, tempers flared, and she gleefully gave them the finger.

With both fists.

But bad behavior in public is no longer without cost. What did the angry bikers do? Naturally, they pulled out their phones and captured a photo of her flipping them off—as well as one of her North Carolina license plate. They posted the story on Facebook, and, within hours, it was shared three thousand times, and her full name and photo began circulating on social media.[25] "She could use a troll," said one commenter. "I hope all her Friends recognize her!!!" said another. "She needs some public shaming," chimed a third.

Shaming someone to prove a point can provide some instant gratification. What happened next, however, was completely scary. The woman's private home address was given out, as well as her work phone number. Before long, this financial analyst was receiving threatening calls at home and work and deactivated her social media accounts. People even began harassing members of her family, who had no control over her actions. (I, personally, would hate to be blamed for the things some members of my own family do—wouldn't you?)

Could Mercedes Mom have mended fences with the cycling community by coming out on social media with a huge mea culpa? Possibly. Who can't relate to having an occasional moment of road rage? But will her online presence forever be associated with this one lapse of driving etiquette? For sure.

WHO'S SHAMING WHO, NOW?

Jennifer Livingston is a working mother of three and a local TV news anchor. One day, a viewer felt compelled to voice his opinion about her weight, sending her this email:

Hi Jennifer,

It's unusual that I see your morning show, but I did so for a very short time today. I was surprised indeed to witness that your physical condition hasn't improved for many years. Surely you don't consider yourself a suitable example for this community's young people, girls in particular. Obesity is one of the worst choices a person can make and one of the most dangerous habits to maintain. I leave you this note hoping that you'll reconsider your responsibility as a local public personality to present and promote a healthy lifestyle.

Like so many who indulge in fat shaming, this viewer wrapped his attack in concerns about the obesity epidemic and its consequences for public health, sweeping aside factors like Jennifer's genetic predisposition to obesity, her battles with weight, and her personal self-acceptance.

Although Jennifer's first instinct was to ignore the criticism, when she saw that he'd called her a poor role model, she felt she had to speak out. Her husband encouraged her, posting the viewer's email on his Facebook page, and Jennifer responded to it during one of her live, televised news broadcasts.[26]

The truth is, I am overweight. You could call me fat. And yes, even obese on a doctor's chart. But to the person who wrote me that letter: Do you think I don't know that? That your cruel words are pointing out something that I don't see?

You don't know me. You are not a friend of mine. You are not a part of my family, and you have admitted that you don't watch this show. So you know nothing about me but what you see on the outside. And I am much more than a number on a scale…

I leave you with this: to all of the children out there who feel lost, who are struggling with your weight, with the color of your skin, your sexual preference, your disability, even the acne on your face, listen to me right now. Do not let your self-worth be defined by bullies. Learn from my experience—that the cruel words of one are nothing compared to the shouts of many.

Probably to his surprise, the viewer, a security guard, found himself on the defense. The Facebook post received more than three thousand comments publicly attacking *him*, in the spirit of this one: "It's so easy for people to sit behind their computer or phone screen and type up a bunch of nastiness aimed at hurting another human. One quick click of the mouse as it hovers over the 'send' button and the deed's done. Do you feel good about yourself then, bully?"

He did later apologize for his message, claiming that he was obese as a child and saying, "I've been fighting all my life."[27] His experience shows that those who start a shaming war never know when the tables will turn and they will find themselves on the receiving end.

Is there a point where shaming the shamer goes too far?

Shaming for Profit

Air travel today seems to bring out the worst in many of us. Who hasn't secretly wanted to scream when your bags are lost, your flight is delayed, or you're separated from sitting with your little ones? One angry mother's 2:00 a.m. tirade toward a gate agent was captured on a cell phone video after a twelve-hour flight delay at LaGuardia Airport. You almost can't blame this mom, who was fearful that her family would miss their pricey Disney cruise, for losing it. One equally upset fellow passenger filmed the blowup and posted it on YouTube, with the title, "A Very Disgruntled AA Passenger," trying to document the passengers' frustration and the flight staff's indifference.

The news media picked up on the story, which quickly morphed into "FURIOUS MOM HAS AIRPORT MELTDOWN AFTER TWELVE-HOUR DELAY RUINS DISNEY CRUISE."[28] Of course, the judging kicked in immediately, as online commenters said that this mother was a bad role model, pathetic, crazy, spoiled, and worse. "Hope her employer sees this and fires her," wrote one.[29]

But take a closer look, and you'll see that this was not your ordinary case of shaming gone viral. Instead, the rapid distribution of this mom's massive meltdown was a deliberate case of shaming for profit. The passenger who filmed the outburst was apparently approached to sell the rights to the video to Jukin Media, a behind-the-scenes middleman that cuts licensing deals to create its next YouTube or morning-show sensation.[30]

Founded in 2009 by Jonathan Skogmo, who at the time was working in TV production, the company is headquartered in Culver City, California, where a team of researchers pores over computer screens each day, hunting for diamonds in the rough. The one-hundred-plus-employee company sweeps in when alerted that a new video seems poised to go viral and cuts a deal with the creator, either buying the copyright outright, or arranging to share profits from ad sales generated on YouTube and permissions fees charged to TV shows like *Ellen, Inside Edition,* or *Today* for airplay. To date, Jukin has thirty thousand videos in its stable, many your typical Internet fare of cat antics, alongside every form of its "fail" imaginable—wheelie fail, backflip fail, zip-line fail. Jukin execs say they have paid out more than $10 million to viral-video owners.[31] But in some cases, the ones profiting from these deals are not necessarily the ones who suffered these moments of humiliation—they're just the people who caught them on video.[32]

What responsibility do those who peddle shame bear? Company executives say that situations like these are "extremely rare," and that Jukin attempts to check whether the filmmaker received permission and will blur out people's faces, if requested. "We make a judgment," says CEO Skogmo. "We definitely feel like we have responsibility, when [cases like this] do come up."[33] But the company also defends its actions, pointing out that it doesn't commission these clips and isn't the first to post them online. Jukin doesn't create the content—it simply provides the megaphone.

Seeds of Shame

Others in the business of shaming wonder whether we have gone too far.

It was late August 2009, and Andrew Kipple, his brother Adam, and their friend Luke Wherry were three single twenty-somethings out on a late-night run to a newly opened Walmart in Myrtle Beach, South Carolina, when they had an epiphany: America has no shortage of crazy, and it can often be found at one of the roughly five thousand Walmart stores nationwide. Where else can you see people shopping in superhero costumes? Or nothing but a towel? Or toting a goat on a leash? And wouldn't it be hilarious to document some of this insanity for everyone to enjoy? The trio returned home and quickly slapped together a website, People of Walmart. With its clever captions and outrageous display of Crazytown, USA, the site soon became a media darling and Internet sensation, drawing seven million hits that fall. "A lot of people have the misconception that we're just a bunch of mean, angry people," Adam Kipple once told his hometown newspaper. "We're not trying to hurt anybody. We're just trying to have some fun."[34]

And it is funny. Hilarious, even. You may want to be morally outraged but find yourself helplessly chuckling at the parade of butt cracks on display. Who knows—who cares?—if the shoppers featured might have felt a little chagrined, over the years.

These days, overshadowed by the clickbait fed to consumers on sites like BuzzFeed, Vulture, and Vice, People of Walmart is markedly less popular than it was in its peak years, when it received millions of visitors per month and fifty daily submissions to choose from. "Our fifteen minutes is definitely coming toward the end," conceded Andrew Kipple when interviewed.[35]

Still, the effects of their empire remain: 1.5 million Facebook followers (until the mega social media platform deleted their page in late 2016, and they had to start over), three spin-off books (disclosure: released by our mutual publisher), calendars, and a spawned network of some thirty like-minded shrines to lampoonery: Neighbor Shame, WTF Tattoos, etc.

"We're amazed it's 2016 and we're still doing it," admits Wherry. The three spend an hour each morning freshening up content, then turn their attention to their day jobs, Three Ring Focus, a ten-person web design and marketing firm they founded. All are now married, two have children, none have any regrets.

What does it feel like, to be one of those who sowed the early seeds of online shame?

"I think people would be surprised to meet us. We're not a bunch of good-looking guys that think everybody [else] is funny," says Kipple. "We're just normal idiots that had a funny joke that went viral." He says they never experienced any negative consequences. "Any time [People of Walmart] did come up, it was always positive," he says. "I can't recall a single time [that] I told someone I [had] created the site and got any sort of rude remark or backlash."

Of course, the three are aware that they have their share of detractors. One woman in Ypsilanti, Michigan, took to her local media in 2011, after finding that her mother was featured on the site decked out in all black, with the caption "A member of the Canadian division of the Trenchcoat Mafia," a reference to the Columbine shooters.

"It upsets me," she told TV news cameras. "We have no privacy shopping? So, I could go into any store and take a picture of anybody or their children and put it up on a web page?"[36] Essentially, yes. A law school grad, Andrew Kipple rests assured that sites like these are on firm legal ground. The entire enterprise basically subscribes to the principle that there is no reasonable expectation of privacy in a commercial public space like a megastore.

The site even freely publishes a running litany of its hate mail. This one, from "Offended" is fairly typical:

> You should be ashamed of yourself. You are nothing but a bully. You should be disappointed in yourself. I cannot fathom how you get joy out of making someone else suffer. Grow a heart, a conscience, and a soul. Learn some morals and values.

"There were a ton of people that would send us emails and tell us, 'We're going to sue [you],'" Andrew Kipple recalls. (None ever have, he says.) "People's reaction was either, this is hilarious—or they absolutely hate us and want to kill us. To this day, that's still something that annoys me. When someone tries to take up the cause for other people that they don't even know. We'd get these emails about what pieces of shit we were, how terrible. How we must think we're all sexy six-foot-two hunks, which couldn't be further from the truth. We would sit there, Luke would call us in, and we'd read them and laugh. Some of the hate mail would leave us in tears."

However, the crew contends that they *do* have standards for what they will and won't post—to a point. The mentally ill, the physically handicapped, beleaguered Walmart employees—all off-limits. Their moral compass can be summed up as follows: Fat guy in a scooter? *Not funny.* Fat guy in a scooter wearing an outrageous hat or a size extra-small T-shirt? *Funny.* "This is supposed to make people laugh—we didn't want to be total jerks," Andrew Kipple says. "Those are the principles we go by." They also claim to have a mechanism for yanking any photo upon request. "If anyone has ever asked us to take [a photo] down for any reason, we take it down. There's no sense getting someone upset over it."

Still, when asked if he feels that the existence of People of Walmart has played a part in the growing culture of public shame, Andrew Kipple grows reflective.

"I'm thirty," he responds. "Maybe it's because I've gotten older, or because times have changed, [but] I can see how people have really gotten mean toward one another, are really about public shaming. I do have to follow that up with, 'Can I just sit here and feel like I've got clean hands?' I don't feel like I'm a big jerk, but at the same time, we were probably part of adding to that genre. So it's tough for me. I can't act like I have clean hands in this. We know who we are as people, and we say we don't do it in a mean-spirited way. I can see someone not in our shoes, saying, 'Dude, who are you kidding? Your website is just making fun of people online.' I have to be open-minded and fair [about] what we contributed to it."

He considers, then adds quietly, "Yeah, I think about it."

CHAPTER 2

NO ONE IS SAFE FROM CYBERHUMILIATION

"Anybody can say anything about anyone today anony-mously online, and there's really nothing you can do about it."

—**Patrick Ambron, CEO of BrandYourself**[1]

If you're like most people, you've probably Googled yourself before. Were you happy with the results? Probably not. A 2012 survey by BrandYourself and Harris Interactive found that nearly half of adults who searched their names online didn't much like what they saw.[2] And if nothing bad pops up now, just consider that all it might take for your respectable reputation to unravel is being in the wrong place at the wrong time. "Take a look on any newsfeed and you'll see that outrage has become a game of sorts," wrote Patrick Ambron, of the online reputation management firm BrandYourself, in the Huffington Post. "What people fail to realize is that in an age of digital permanency, today's shame is stickier than molasses and far more toxic."[3]

One of Ambron's clients told this story: "I made the dumb mistake of getting involved with a less-than-stable older woman when I was in my mid-twenties. At one point, I had to call the cops because she was showing up at my apartment after having told her explicitly that she was not welcome. She ended up posting every picture of us she could on Flickr and refuses to take them down. I've spent years in sports, philanthropy, and travel, and yet when you search my name, you get a bunch of pictures of me and some woman I never want to have contact with ever again in my life. It's like this skeleton in my closet. And of course, yes, women have Googled my name and found them. It's a bit awkward and handicapping to say the least."[4]

Rich Matta, the CEO of ReputationDefender, says that the number of clients approaching his reputation management firm for help has spiked in the decade since the business was founded. "We're seeing an enormous rise [in] online attacks for three main reasons," he says. "First, attackers now understand that online attacks have significant potential to negatively impact their intended targets—and they're growing more sophisticated, selecting approaches that maximize damage. Second, social media is massively pervasive—it's a 24/7 opportunity to defame yourself inadvertently or be shamed and defamed by others. Finally, there's financial incentive to participate in the online marketplace for defamation schemes: fake review sites, mug-shot sites,* certain legal records sites, webcam blackmail, and more."[5]

Let's explore some of the more extreme examples of digital disasters that can strike any one of us before we turn to learning how we can prevent and survive them.

Nonconsensual Porn and Revenge Porn

Sexting might be considered the new form of flirting, but that doesn't mean a sext isn't going to get you in trouble, if your recipient decides to use it for unsavory purposes. Annmarie Chiarini, for example, thought she was innocently accepting a Facebook friend request from an old high school friend. How wrong she was.[6]

Although it started out as a typical romance, this flame quickly turned into fire. After her boyfriend begged her for weeks for nude photos, Annmarie begrudgingly gave in *slightly*. She wouldn't outright pose nude but allowed some seminudes. It wasn't long before their relationship starting going south. He became extremely possessive and verbally abusive, accusing her of infidelity. After seven months of their reunion, she called it off.

Almost immediately, the threats started. She was told she'd better take him back—or else a CD of her nude photos would be up for auction on eBay.

Annmarie was not only a mother of two, she was also a

---

* Websites that post mug shot photos of people who have been arrested and charge a fee for removal.

fortysomething English professor at a college in Maryland; she had a reputation to protect. But it wasn't long before the threat became a reality. Her ex posted the photos in an online eBay auction, reposting them as fast as she could request that the site pull them down.

"I felt very helpless," she recalled in an interview. "Someone else was in control of my life, someone else was shaping public perception of me. I could no longer define who I was and how I wanted people to see me; someone else was doing that. He was using our intimacy and our sexuality to shame me, and play on public opinion that women who are sexual are 'less than,' are not good, upstanding people, are not worthy of positions in academia and parenthood. There was this fear, this raw, all-consuming fear, of what was going to happen, what the backlash was going to be. It was terror, because there's so much social collateral that goes along with being a woman who has chosen to be sexual. If any social norms are violated, there is an angry backlash. There's something about a woman's sexuality that makes people angry."[7]

Links to the nude eBay auctions were showing up on different platforms, including the Facebook page of the college she worked at, where she knew students and colleagues saw them. The fear of what they would think, and the potential judgment of her friends and family, was overwhelming. She recounted the thoughts running through her mind: "They're going to shun me, I'm going to lose my job, they're not going to want to be my friend, they're not going to feel my pain or understand that this is scary."

Annmarie attempted to get help, first through the local

authorities, which was a dead-end. "The police were completely dismissive," she said, describing their attitude as, "It was my fault, I was stupid, there was nothing they could do, I should have known better." There were simply no laws in place to help protect her against this type of online behavior.

She found herself waking in the middle of the night, running a Google search on her name, checking eBay and Facebook repeatedly, until she was able to settle back down and return to sleep.

The last straw came a year later, when an online profile was created impersonating her, with her nude images and invitations to men to come to her home to have sex with her. It read: HOT FOR TEACHER? WELL, COME GET IT! "I realized I was an 'off-line' target," she said. "My full name, [the] town where I lived, the solicitation for sex was posted on the profile, so people could find me." Now she was physically vulnerable—and scared for her safety.

She reached out again to state police and the Baltimore division of the FBI for help, and at her lowest point, she contemplated taking her own life. The site was up for fourteen days and had received three thousand views before a tech-savvy friend was able to help her get her pictures removed. "That's nothing," she said in retrospect, comparing herself to other victims of revenge porn she has since worked with, some of whose images have spread to some three thousand different websites. "I was phenomenally lucky."

Annmarie would eventually turn her anger to activism and was instrumental in persuading the state of Maryland to enact a law criminalizing nonconsensual porn.

☞ The takeaway: Even adults in a committed relationship can get tangled up in a sexting nightmare.

Noods, Slut Pages, and Sexting Scandals

In the affluent seaside town of Duxbury, Massachusetts, (nicknamed "Deluxeberry," for its oceanfront homes) rumors of the not-so-secret Dropbox account had been circulating all year. On it were said to be folders named after some fifty Duxbury High School girls, each containing revealing or even nude photos.

How had those photos, known as *noods* in Internet slang, come to exist? Some girls had sent their photos to boyfriends, trusting that they would keep them private; others, to boys they were crushing on, hoping to impress or land a potential hookup. Some had refused to comply, so out of frustration or vindictiveness, someone doctored up a nude photo with their name or face and posted it anyway.

Eventually, school authorities were tipped off. "People were getting called down to the office left and right," one female Duxbury High student, "Ginny," explained in an interview. "Girls were freaking out, and the boys were deleting the Dropbox [app] off their phones. A lot of the girls who knew they were in the Dropbox weren't coming to school."[8] Boys and girls were hauled into the police station for questioning, and TV news crews parked themselves on campus for five days, trying to convince students to speak with reporters.

"There are some young people here who are very

embarrassed and very upset," police chief Matthew Clancy observed at the time in the hometown paper.[9] But when you talk to the girls directly, you find that what they dreaded most was not the reality that their naked bodies were circulating online, but the news getting out to the adults. "I don't think the girls were all that embarrassed," Ginny says. "They just were afraid of getting in trouble and sent to the police station with their parents."

Although the police emphasized that they viewed the girls as the victims and did not intend to charge them with a crime, there was still cause for concern. Because Massachusetts, like many states, had no specific laws related to minors sexting, any teen girl caught distributing a nude photo of herself could technically be charged with a felony count of child pornography, bringing prison time and placement on the sex offenders registration list.

And yet, in a twisted bit of irony, many girls were secretly pleased to be on the list. Getting placed on the Dropbox was considered a coup. "It was a bit of a beauty contest," says the mother of one girl. "Some are mortified, some are proud. There's a lot of brazenness."[10]

"Girls were happy that their pictures were put there," agrees Ginny. "It made them feel like all the boys loved them."

If you think this is an isolated case, think again. As documented in Nancy Jo Sales's *American Girls: Social Media and the Secret Lives of Teenagers*, this phenomenon of boys pressuring girls to share nudes, and then often betraying that trust by sharing them on so-called slut pages, has become a rite of passage, popping up in small towns like Cañon City, Colorado;

Winnetka, Illinois; and Knoxville, Iowa.[11] "Every one of my colleagues have been dealing with this; it's becoming a norm," explains Police Chief Clancy. "No one is taking it seriously."[12]

One study of teenagers in Texas basically concluded that sexting is the new first base.[13] In a 2012 survey of college students published by Elizabeth Englander of the Massachusetts Aggression Reduction Center, 30 percent admitted that, during high school, they had sent nude photos, and 45 percent reported that they had received them.[14] The risk of those photos getting passed around is not insignificant—the students said that about a quarter of the time, their photos were spread beyond the original receiver, especially when they'd been pressured to send them in the first place. In a 2015 follow-up, Englander found that 70 percent of the sexters reported feeling coerced, at times, to do so.[15]

"I'm asked all the time for pictures," remarks Ginny, the Duxbury junior. "I've experienced boys begging and begging... Boys always say, 'Of course I won't save it... I won't show anybody...' But really, they do. Even if you send it over Snapchat, the boys have an app that will save the pictures without the girls knowing... Stuff like this happens at other schools. But I've never seen it [be] taken to the extent that Duxbury takes it. They make it so unbelievably humiliating."

"It's in every school," counters longtime Duxbury High School principal Andrew Stephens, a concerned father with two daughters of his own. "The browbeating that goes on with young men to young women to get these is ridiculous, until they say, 'Here, fine, just shut up.' The ability to stand up and say no, and be willing to have whatever is threatened [actually] be done, that's a tall order to a lot of girls." Adults, he says, need to accept that this is the new reality. "I've talked to parents that are stunned clueless about what is happening. I live it. If you don't understand this world and are giving advice based on a 1985 model, that doesn't work anymore."[16]

What is the larger message being sent here? On top of being shamed and often betrayed, these girls were also subjected to the typical blame-the-victim message. "I want our young women to understand that they need to think better of themselves," scolded the school superintendent, Dr. Ben Tantillo.[17] Ginny bristles at this message. "Girls don't need to be told to 'think better of themselves,'" she responds. "Boys need to stop being immature and sick creatures, grow up, and not post girls all over a site for everyone to see. Why are they not telling the boys to stop hounding the girls for naked pictures?" Meanwhile, the boys who merely passed photos along seem to have gotten off relatively scot-free, receiving only this mild scolding from Tantillo: "We want our boys to know that this is not an appropriate way to treat young women."

As for the person or persons who created the account? As of March 2017, the police have identified two to three of the boys that created the Dropbox, and they have been served warrants, although we are still waiting to see if charges will be filed. They did note that the boys are being cooperative and are remorseful.

The Duxbury mom believes that the town's teens will be more cautious about what they put out there in the future.

"Talk about learning a lesson. I would say the entire high school learned the lesson."

But have they? Most disturbingly, these girls seem to have learned something else: how to preserve their anonymity in an attempt to circumvent shame. According to Ginny, girls are still sending nudes to cater to boys' demands—but leaving their faces out of the frame, to have deniability. "A lot of the girls said 'that's not me' to the pictures that had no faces, to save themselves— even when it *was* them," she explains. "There was no way to really prove it." As for herself, Ginny is left wondering what is out there and what might come back to haunt her down the line. "I did hear my name was on that Dropbox. I did hear my pictures were in there," she admits. "I have never sent anything with my face—I don't know for sure." Sadly, investigators say these photos linked to their names, true or forged, headless or not, will likely trail these girls long past graduation day.

☞ The takeaway: Handle sex and tech with care; noods gone viral are impossible to take back.

Sextortion

Ashley Reynolds was only fourteen, a bubbly high school freshman sitting on her aunt's porch in Phoenix, Arizona, one hot summer day, when her phone buzzed with a text: I have naked pictures of you. Ashley read the words but was sure she had never sent anyone such photos. "I was just like, 'Ew,' and ignored it," she told *Glamour* magazine.[18] But the messages wouldn't stop. For the next few hours, her phone buzzed. And buzzed. And buzzed.

The anonymous stalker threatened to send the photos to her friends and family if she didn't send him another. Ashley was sure he was bluffing, but she was also freaked out—how did this stranger even get her cell phone number? Eventually, she caved, hoping a photo of herself in her bra would satisfy him. "I thought he would go away," she recalled. But of course, he didn't. Within days, he came back, this time with increased pressure for more explicit pictures—and now, leverage over her.

This behavior is known as *sextortion*.

For months, Ashley struggled to keep her secret and meet her extortionist's increasingly twisted demands, as he hounded her for up to sixty photos per night. Eventually, Ashley's mom stumbled upon an email message and the images on their home computer and forced Ashley to confess what was happening. Together, they contacted the National Center for Missing & Exploited Children (www.missingkids.org), which recommended that they cut off all contact and launched an investigation. Unfortunately, Ashley's attacker was highly tech savvy, hiding his IP address using proxy servers, which made it impossible for investigators to locate him.

Angered, he made good on his threats, posting Ashley's revealing photos and sending the link to many of her friends on social media.

Ashley, of course, was devastated.

However, in doing so, he'd slipped up. This time, police were able to trace his IP address to his home, arriving there to find a twenty-six-year-old man, a computer with more than eighty thousand images, and evidence that he'd been doing the same thing to some *350 girls nationwide*.[19]

The FBI's advice on sextortion is basic and clear: never provide predators with easy ammunition. In other words, don't do things like flash someone on a video chat service like Skype or send a photo at a stranger's demand. It's also important to realize that these cybercriminals aren't always strangers—there are stories of teen girls being sextorted for images by boys they know personally. And the victims are not always young women—males can be preyed upon too, often with a motive of financial extortion.

But most disturbingly, you don't even have to provide someone with a nude photo to find yourself exposed against your will.

For Cassidy Wolf, crowned Miss Teen USA in 2013, the first sign that something was wrong was when she started getting notifications from her social media accounts that someone had tried to change her password. Thirty minutes later, she received an email from a hacker, saying that he had hijacked her computer's webcam and taken nude photos of her while she was undressing in her bedroom. "I wasn't aware that somebody was watching me [on my webcam]," she told the *Today* show. "The light [on the camera] didn't even go on, so I had no idea."[20]

In exchange for not making the embarrassing images public, the hacker made a number of demands of Cassidy, as well as dozens of other women he'd victimized: that they send him more nude photos or good quality videos, or perform private, live Skype sex shows.

This individual turned out not to be some creepy, far-flung stranger, but someone who Cassidy knew: nineteen-year-old Jared James Abrahams was a former classmate from her high school days at Great Oak High School in Temecula, California. He was eventually caught, pleading guilty to charges of extortion and unauthorized use of a computer,* and sentenced to eighteen months in jail.

According to the FBI, cases of sextortion are alarmingly on the rise. A 2016 Brookings Institution report documented seventy-eight recent cases, likely involving three thousand victims—and possibly even thousands more.[21] All the perpetrators were male; one worked at the U.S. Embassy in London, another was an intern for a Republican presidential campaign, and another lured girls by pretending to be pop star Justin Bieber. "Sextortion is brutal," the researchers concluded. "Many cases result, after all, in images permanently on the Internet on multiple child pornography sites following extended periods of coercion."[22] It's also important to note that victims are getting younger and younger, as parents are ignoring age restrictions and recommendations set by the sites, and kids are finding ways to get on social platforms before the required age. A nine-year-old girl in California became a victim of a twenty-four-year-old male posing as Justin Bieber in Massachusetts. This perpetrator

---

* As defined by §156.05 of the New York Penal Law, unauthorized use of a computer occurs when an individual accesses a computer without permission to do so and the system is protected in a manner to prevent unauthorized use.

is expected to face charges including extortion, manufacturing child pornography, and communicating with a minor with the intent to commit a sex act.[23]

Although we frequently hear about women as victims, men are not immune. At only twenty-one years old, Jake Curtis ended his life after meeting a girl online and becoming her victim of sextortion. She demanded money, or she would destroy him and hurt his family, she claimed. According to his mother, he decided suicide was his only option, leaving a note that outlined the details of this horrid experience.[24]

Ashley Reynolds has bravely agreed to be the public face of this story, hoping to reach the hundreds of girls who were also victims of her tormenter but might still be unaware that he was caught and convicted. She hopes that speaking out can help prevent other girls from a similar fate, despite the humiliation that still lingers with her. "I do occasionally walk down the street and wonder if anyone has seen [the photos of me]," Ashley told *Glamour*. "And I definitely have trust issues."[25]

☞ The takeaway: Sextortionists can target all types, men and women, adults and teens, famous and otherwise. And you don't even have to send a nude to become a victim.

ACCIDENTAL SHAMING

With the popularity of drones on the rise, there's a new risk to consider from the eye in the sky—even when you're lounging in the privacy of your own backyard. One Australian grandmother was sunbathing topless when a realtor's drone snapped an aerial photo of her neighbor's property, in order to list the house for sale. Later she discovered, to her chagrin, that she was inadvertently included in the shot, and her naked body was now plastered on billboards around town. "It's in the real estate magazine, it's on the Internet, and on the [bill]board, and I'm really embarrassed," the distressed woman told one newspaper.

Thankfully, she caught one lucky break: she happened to be lying facedown![26]

Ugly Polls

Once upon a time, a teen wanting to be mean might tear some paper from a binder, scribble something nasty about a classmate, and pass it around school, sending it into the rumor mill for a week or so. Today, these slams are posted online for all the world to see, and they have the potential to inflict much more harm on their victim's psyche.

It was a cold winter Tuesday when an anonymous poll was the talk of the school at Holy Trinity High in the small town of Torbay on the island of Newfoundland in Canada; the poll ranked which girls in the senior class were the prettiest—and the ugliest. Lynelle Cantwell, then a seventeen-year-old senior, had just arrived in math class that day when she first heard classmates discussing "The Poll," which was posted on the website ASKfm. Although she wasn't familiar with the site, it is generally known for its anonymous, and often vile, questions that no one would dare ask in real life.

The poll was called "Ugliest girls in grade 12 at HTH," an abbreviation for her school, and ranked each girl by the percentage of people who had voted for her.

And Lynelle's name was in fourth place.

"It hurt. A lot," she later told the *Toronto Star*.[27] "I'm not going to lie. It kind of tore me up inside a little bit. I was like, 'Oh, my God, like why would somebody be so mean as to post stuff like this?'"

What makes this story unique is that Lynelle found a way to rise above these small-minded voters. "It outraged me," she told CBC News.[28] "All of these girls are hurt because of this one person who has no heart or sympathy for anything."

Here's what she posted on Facebook in response.

To the person that made the "ugliest girls in grade 12 at hth" ASK.fm straw poll. I'm sorry that your life is so miserable that you have to try to bring others down. To the 12 people that voted for me, to bring me to 4th place. I'm sorry for you too. I'm sorry that you don't get to know me as a person. I know that I'm not the prettiest thing to look at. I know I have a double chin and I fit in XL clothes. I know I don't have the perfect smile or the perfect face. But I'm sorry for you. Not myself. I'm sorry that you get amusement out of making people feel like shit. I'm sorry that you'll never get the chance to know the kind of person I am. I may not look okay on the outside. But I'm funny, nice, kind, down to earth, not judgmental, accepting, helpful, and I'm super easy to talk to. That's the same for every other girl on that list that you all put down. Just because we don't look perfect on the outside does not mean we are ugly. If that's your idea of ugly then I feel sorry for you. Like seriously? Get a life.[29]

"When I posted it, I expected a comment from my aunts and uncles, acquaintances on Facebook, maybe a couple dozen likes," Lynelle recalled in an interview. "Then I went to gym class and put the phone away, and [when I got back,] my phone had gone crazy. I never ever in a million years thought it would go as far as it did."[30]

Lynelle's post received hundreds of messages of support, stating the obvious: the only ugly people in this scenario were the ones who participated in the poll. Her words were shared more than eight thousand times, and she was feted by the news media in Canada and beyond. In the following months, Lynelle would celebrate Pink Shirt Day to fight bullying, jet off to Toronto as an invited guest at a four-day youth leadership summit, and attend her school prom, a formal dinner and dance followed by zip-lining.[31]

Was this ugly poll the first of its kind to circulate online? Of course not. Lynelle's school district reported that five other schools in their province alone had dealt with similar ugly-girl polls. Only a month later, the same kind of poll popped up in a school in Port aux Basques, on the opposite side of the island. "It made me upset," Lynelle said, "because I felt my message did not work at all; they're doing it again."

But months after the story had died down, Lynelle received an explanation that she never saw coming. "I got a private message from the person that started it," she said. "They apologized and explained their reasoning. It was a nice apology. I appreciated it."

The person behind the ugly poll wasn't a group of mean boys ranking girls' sex appeal.

It was a fellow female classmate.

Big-hearted Lynelle was able to do what most of us would not: forgive her onetime nemesis. "We're good," she said simply, explaining that the girl's motivation had been, more than anything, high school cliquishness. Her advice to others being cybershamed? "Try to not stoop down to the level of whoever is picking on you," advised Lynelle, now a high school graduate aiming for a future career in the Royal Canadian Mounted Police. "Stand up for yourself—but take a higher road."[32]

Mean Memes

Ashley VanPevenage, a twenty-year-old college student from Tacoma, Washington, was self-conscious about attending a friend's upcoming twenty-first birthday party. Like many of us, Ashley had been struggling with acne, but a recent allergic reaction to a medication had left her face looking worse than ever. She asked a friend from church who was an aspiring makeup artist to apply heavy concealer to help hide her blemishes. To demonstrate her skills with a makeup brush, Ashley's friend took a before-and-after photo for her "MakeupByDreigh" Instagram page.[33]

The page already featured many other similar shots, but this particular photo was repurposed against both girls' wishes. Ashley discovered that a Twitter user had lifted the photo from the page and placed it on a photo-sharing site, Mobypicture, with the quip, "The reason why you gotta take a bitch swimming on the first date." Another Twitter follower chimed in: "I don't understand how people can do this and I can't figure out how to conceal a single pimple on my face," which was picked up by a satire account with five million followers, @SoDamnTrue.[34] Soon, the cruel meme was circling Facebook, iFunny.com, and beyond, with Facebook strangers from around the world posting things like saying they're sorry for the guy who wakes up next to her. Ultimately, the image was shared four million times, and Ashley found her self-confidence taking a nosedive.

Deciding to take action, Ashley uploaded a video to YouTube called "My Response to My Viral Meme," which has been viewed 1.6 million times and counting. She told viewers, "It doesn't matter what's on the outside of your face—the only thing that matters is who you really are."[35]

Ashley also created the antibullying campaign #CureTheHate, devoting herself to spreading a message that we all need to keep in mind: there is often a real person on the other end of these mean memes who may not find it quite so funny.

Elder Shaming

It was June 2012 when the nation got a taste of the cruelty toward a sixty-eight-year-old grandmother who was maliciously taunted by four seventh graders on a school bus. Former school bus driver Karen Klein was working as a school bus monitor in upstate New York for the Greece Central School District when this infamous incident was deliberately captured on video by her tormentors.

For ten minutes, the boys harass and bully Karen relentlessly about her age, her appearance, and even the tears she tries to wipe away from her eyes as she tells them, "I'm crying." They call her "fat-ass," poke her, taunt her about her love for Twinkies, and threaten to egg her house, to urinate on her door, and to stab her.

One of the harshest taunts was, "You don't have a family because they all killed themselves because they don't want to be near you."

Karen's oldest son had committed suicide more than a decade ago.

This would have been awful enough to endure for just one afternoon. But the boys went further. They videotaped the entire episode and uploaded it to Facebook and then to YouTube with the title "Making the Bus Monitor Cry." The audacity of these boys ensured that the story was quickly picked up not only by the local news, but by national media as well, and the video received millions of viewers, many of whom were angered enough to send the boys death threats. (Ultimately, they were punished, suspended from school for a year, and forced to perform community service.)

While the video was burning up social media, observer Max
Sidorov had enough. He decided that Karen deserved a break
and set up an online collection to send her on vacation. The
response was overwhelming. More than thirty-two thousand
people from eighty-four countries pledged more than $703,000
in donations.[36] What did Karen do? She started her own cyberbullying foundation.

☞ The takeaway: Our failure to instill empathy in young people has
created a culture of cruelty.

SHAMING THE HELPLESS

Grandma may not even own a smartphone, but that doesn't mean
she's safe from being cybershamed. An updated 2017 ProPublica
investigation reported on some sixty-five recent cases of nursing-
home staff sharing photos and videos of their geriatric residents,
often in humiliating or compromising moments.[37] At one facility
in Michigan, a nursing assistant was accused of snapping a photo
of an elderly woman on the toilet and sharing it on Snapchat.
In a similar case, an Indiana nursing assistant shared a photo of
a senior's behind on Snapchat. In San Diego, an employee was
charged with elder abuse for sharing a partially naked photo of a
patient getting into the shower. "It's taking advantage of the weak,
and it's sick," said one outraged relative, voicing the thoughts of
many. [38] How would you feel if this happened to your own mother
or grandmother?

Body Shaming

*"I thought I was safe. But the truth is, no one is safe.
There is no such thing as bully-proofing. This could
happen to anyone."*

—Galit Breen[39]

In June 2014, on her twelfth wedding anniversary, Galit Breen,
a Minnesota mommy blogger and mother of three, decided to
write a listicle for the Huffington Post titled "12 Secrets Happily
Married Women Know."[40] Along with the article, she included a
few photos of herself on her wedding day. Galit had never before
publicly posted the photos of her younger, heavier self. But she
felt that these wedding day shots portrayed her happiness, the
glowing bride and groom, so she sent them along to her editor.

Galit knew "never to read the comments," but she checked in
on HuffPo's Facebook page anyway, to see the reaction her words
of wisdom were getting. Once she did, she couldn't look away,
refreshing the page over and over.

"One thing you didn't learn is 'don't marry a heifer,'" lobbed
one critic.

"WE GET IT!" commented another. "Huffnpuff…you love
fat women…we get it…enough is enough."

"People weren't commenting about marriage or weddings
or my article or my writing," Galit recalled sadly in an interview.[41] "What they were commenting on is how fat I looked
in my wedding dress… When I read those words, I was really

devastated. I only showed them to my husband, because I was also ashamed and embarrassed. But because I didn't tell anyone, I was also very much alone. I cut myself off from both my online and in-person support systems. I couldn't get past the shame of it, the embarrassment. I think I was depressed. It took me a few months to pull myself out."

Eventually, Galit realized that she had a choice: "Keep on being depressed, or find a way to speak up."

That fall, Galit penned a response piece for xoJane.com titled "It Happened to Me: I Wrote an Article about Marriage, and All Anyone Noticed Is That I'm Fat," which quickly went viral, landing her on the *Today* show and in *Time* magazine.[42] You might think that all this attention would cause her critics to feel some remorse. But no. None of her fat shamers ever reached out to apologize. Galit was criticized for leaving their full names in her article, forcing them to turn their Facebook accounts private, as they in turn were harassed. Today, she is unrepentant for that choice. "They made a decision: to log into Facebook and say awful things," Galit pointed out. "Own it. If you don't want to be portrayed badly, don't act badly."[43]

Teacher Shaming

Ms. B., a fourth-grade paraprofessional at an elementary school in Atlanta, Georgia, loves working with kids and has even been awarded Educator of the Month by her school district. But none of that mattered after she posted on Instagram shots of herself in the classroom, smiling in her body-hugging, knee-length dresses or in jeans and a T-shirt paired with high heels. Parents began flooding Twitter with the hashtag #teacherbae, charging her with inappropriate choice of work attire.[44] Completely disregarding how much the kids loved her or what they learned from her, parents wanted to judge only her appearance—not what was on the inside. And what were these parents saying?

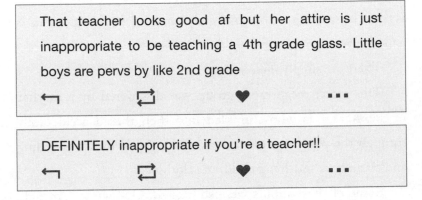

What was really going on? This teacher's aide is a gorgeous fashionista, and her private Instagram account shows plenty of cleavage-baring shots. However, none of her workday clothing was actually revealing at all—often, she was wearing a simple cardigan sweater. It seemed that the only true objection was to her naturally voluptuous figure.[45] After she was forced into the spotlight, the school district reprimanded her, reminding her of its dress code and social media policy.[46]

☞ The takeaway: There is always someone insecure enough in their own body to mock yours.

Baby Bashing

Just when you think you've seen it all on social media, you stumble across a moms' group in Palm Beach Gardens, Florida, set up to intentionally mock other people's babies.

One mom wrote about a toddler's picture, "Before I address this… It… I want to point out that it makes my heart happy that you have a *Mean Girls* tab in your computer. Good stuff. Now, #1, is this a he or a she… You absolutely can not [*sic*] fix ugly. This is a God-given example of such."[47]

There are simply *no words*.

This secret mean-mom group was discovered by a mother in South Florida who stumbled into their thread accidentally through the page of a private Facebook group that was selling and trading second-hand children's clothing.

Some of these moms were so excited. One wrote, "An ugly baby thread… I have died and gone to heaven… Why can't you guys live near me so we can do this over cocktails?"

It's so cruel, it really takes your breath away.

One mother attempted to shame them for mocking her toddler by posting a picture of her as a newborn in the hospital's intensive care unit. She wrote, "This is MY DAUGHTER who was made fun of because she is delayed… So funny, huh? Sick bitches."

Did these cruel moms back down in shame? Not one bit. Instead, these "free-speech advocates" cited the First Amendment and continued right on with their taunting: "This is Facebook, not the Salem witch hunt… This is a free country and I was laughing because it was funny... Don't attack me because I laugh at something! You don't get to control that. Thanks for your comments. Next!"

Free as their speech may have been, shortly after it was exposed, the mean-mom group deactivated their secret thread. Does this mean it isn't happening again someplace else? What's your guess?

————

Trolls will be trolls, but nothing is worse than insulting those who have barely learned to speak. As one writer lamented, referring to a toddler who was fat shamed on Reddit, "This baby doesn't even know how to say the word 'fat' yet. She's been alive for less than two years and even she can't escape the stupid body standards girls face. It's not fair."[48]

In another well-known case, Jessica Benton created an Instagram page for her adorable ten-month-old son, Landon Lee, in October 2015. Her intention was simply to dress him cutely and share the photos. She was conscious of the fact that people overshare on Facebook and wanted to avoid it on that platform, so she turned to Instagram, where she didn't have as many followers. But she didn't expect what happened next. One follower took one of Landon's photos and posted it on Facebook with the caption STUFF CURRY, encouraging his followers to engage in fat-shaming comments about the little boy. It wasn't long before the photo was viewed and shared more than nine thousand times.

Jessica wasn't familiar with sports, so initially, she didn't realize that they were referring to Landon's likeness to basketball star Stephen ("Steph") Curry, twisting it with an insult about his chubbiness. After using Google to find out more, she decided to use the Internet to turn this shaming around.

She told ESPN, "I wanted to really turn it into something good and take control of it and say, 'Okay, we're gonna own this name. Yeah, we're Stuff Curry. We look like the famous basketball player.'"[49]

What people didn't know about the family is that when Jessica was pregnant with Landon, she lost her twenty-year-old son to suicide. His death has inspired her to run @babylandonlee on Instagram, with love and tremendous savvy. She recalled how her late son had told her about being bullied in high school, but she hadn't known how much those episodes of bullying had affected his self-esteem. "I can't say [his death was] directly due to bullying or anything, but I have one kid who's not here with me who told me that people made fun of him. I'm not going to have another kid think that the whole world was laughing at him," she said.[50] Now, nobody's laughing at Landon, who has reached eighty thousand Instagram followers with his antibullying message.[51] If anything, we're joyously laughing along with adorable baby Stuff and his awesome mother Jessica.

☞ The takeaway: Shaming has sunk to new lows when we attack those who don't yet have a voice with which to defend themselves.

Parent Shaming

With a simple visit to the zoo, your life can change forever. On May 28, 2016, the world nearly crumbled for one mother after her three-year-old wiggled past her and climbed into the gorilla enclosure at the Cincinnati Zoo.

Thankfully, her son survived. Tragically, Harambe the gorilla had to be shot. Because a smartphone captured this entire event, it ignited a firestorm among animal-rights activists online. Critics and fellow parents alike were quick to blame this mom on social media for not doing her job as a parent, even sending her death threats. It was heart-wrenching and horrifying at the same time.

Moms have been judging other moms for decades—but now, thanks to technology, our worst parenting moments can become immortalized for all of eternity. We live in a world where there are video cameras on every corner and someone always standing ready with their smartphone, waiting to capture your complete meltdown.

Most parents understand that a toddler can get away from you in a matter of seconds. *Seconds*. So why do we play this mental game, blaming other parents when we so easily could have been in their shoes? Psychologists say we do it to try to prevent something similar from ever happening to us. In other words, much of this blame game is a defense mechanism, a hedge against the inherent randomness of the universe. "We like to think the world is orderly and makes sense, that if bad things happen it's because people are bad or have done something to deserve it or are bad mothers," says Dr. Robi Ludwig. "Sometimes, bad things happen to good people, and that's a scary idea. Nobody wants to think that way—it's too unsettling."[52]

Patrick Ambron, BrandYourself's CEO, wrote a sympathetic piece describing the repercussions that the shaming around the Harambe tragedy would have on this woman's life. "Imagine being the mother, who incidentally works as a day care administrator, going to job interviews for the rest of her life knowing that search results have branded her the worst mom in the world," he wrote, adding that the destruction extended beyond this particular woman. "Several women with the same name as the mother even changed their Facebook profile images to avoid being targeted. These women, who have zero connection to the incident, will also be associated with the incident via Google indefinitely."[53]

It was also the summer of 2016 when one Nebraska family traveled to sunny Florida for vacation. Their Disney World vacation was cut short when their two-year-old son was snatched by an alligator in shallow, beachfront water in a man-made lagoon at their Disney resort.

Sadly, hours later, the boy's body was found.

The blaming quickly began. *Why* did the parents let him wade into the lagoon when signs indicated no swimming? (The signage was ambiguous and did not mention alligators.) How *far* were they from their toddler son? (Within arm's reach.) Didn't they *know* that this was a foreseeable outcome? (Actually, this had never happened before in the park's forty-five-year history.)

A blogger in Dade City, Florida, named Melissa Fenton heard such comments on a local talk-radio show and immediately took to social media, quickly scribbling a post that pleaded with people who were posting online to reconsider their words. [54]

> Parents, I beg of you, stop blaming and shaming other parents...
>
> At the funeral for this two-year-old-boy who died in front of his parents, can you do me a favor? Can you walk up to the mother and say the words that you just typed out last week? Can you?
>
> Can you greet her, hug her, shake the father's hand, and then say, 'Who was watching that little boy? You should have known better. I would never let that happen to MY child.'
>
> Can you do that for me? I mean, you felt those words so deeply in your heart and soul that you typed them for a million people to read.
>
> Certainly, you can say it straight to the faces of the people you meant it for, right?

Fenton's post literally *demanded* compassion for the family.

> Here, let me help you.
>
> Put away your pitchfork for a moment and try this.
>
> To the mother and father who went for a walk on vacation for the last time with their little boy yesterday, I am deeply sorry that you had to experience the worst kind of tragedy possible, an accident. I grieve with you. Your baby was my baby. Your son was my son. I have nothing but love for you, love to help you get though the pain yesterday, today, and

for what is gonna seem like a thousand tomorrows. I wrap my
thoughts and prayers around your aching heart and soul. May
the God of this universe in some miraculous way bring peace
to you and your family.

That is what you say. THAT. And just THAT.

The response was overwhelming. Within hours, Melissa's
words were shared nearly half a million times. As of this writing,
the post has been viewed fifty million times, and there are forty-
five thousand comments and counting. More than any earlier
examples of moments of shame, this plea for empathy seemed to
resonate the most.

☞ The takeaway: Don't let know-it-alls tell you otherwise—there is
no such thing as perfect parenting.

Could we secretly be wanting a tamer digital culture? Aren't
we all tired of the shaming and blaming? I believe many of us
are—however, as we will see in the upcoming chapter, that doesn't
prevent some from committing cyberblunders that could easily
land them in the hot seat.

CHAPTER 3

I CAN'T BELIEVE THEY POSTED THAT!

TWEET REGRETS AND OTHER MISPOSTS

"I never thought that one tweet would ruin my life."

—Suey Park, social activist[1]

Why do we post things that we know could get us in trouble? Are we not thinking it through in the heat of the moment, or do we think no one is paying attention? Are we simply naive, thinking that what we say is only among friends? Or are we the opposite, craving the approval of all those likes or retweets? As we will see, so many times, these messes are entirely of our own making. Your online behavior should be the best reflection of who you are off-line, but so many of us don't live up to that ideal.

It had been a long year for Mrs. H., a popular history teacher at a California high school. She was staying on to teach summer school, enduring her most challenging students all over again. Here's what she blurted out on Twitter one day in late June.

I already wanna stab some kids. Is that bad? 19 more days.

A joke in poor taste? Maybe. But it got worse. Scrolling back through her Twitter feed showed that this wasn't the first comment she'd made that, to some, crossed the line. "I am getting Starbucks for sure before school tomorrow," she also wrote that day. "That way I'll be refreshed AND have something to dump on the little a-holes." And earlier, on May 5: "The problem w/ teaching summer school is I'm gonna get all the [expletive] who failed my class, 2 of whom I wish would get hit by a car."

So happy to be done w/school for 10 days, but especially to be away from the ones who truly try my patience & make my trigger finger itchy.

Y'all think you can hate us teachers? Guess what, we feel the same about some of you. We're just not allowed to show it. #ISecretlyHateYou 😄

> Ya know what sucks the most about teaching summer school? The kids I can't stand, who failed, will be in that class. #SummerRuined

In the wake of tragedies like Columbine and Sandy Hook, "itchy trigger fingers" and stabbing students are not subjects that should ever be joked about. The school's principal was alerted by one of Mrs. H's colleagues, which in turn triggered a police investigation, a heated school board meeting, and an examination of her union contract. Mrs. H. later apologized for the tweets and deleted her account, vowing to avoid social media in the future. "I never expected anyone would take me seriously," she later told the *Mercury News*. "If I had thought for one moment that someone would read anything I said on Twitter and take me seriously, you'd better believe I would have been much more careful with what [I'd] said."[2]

We all know where this teacher is coming from, but sadly, it's no excuse. It's hard to believe that this twentysomething digital native didn't know that what goes online stays online. Mrs. H. was given a written reprimand but ultimately was allowed to continue teaching. But others may not be so lucky. Next time, don't post your workplace gripes on social media. Grab a friend after hours and share it during a "whine and wine" session.

☞ The takeaway: Don't air your workplace woes.

Take as another example the evolutionary psychologist who was a tenured professor at the University of Mexico and a visiting professor at New York University. He had published several books on sex and mating. On June 2, 2013, he felt the need to tweet this:

> Dear obese PhD applicants: if you didn't have the willpower to stop eating carbs, you won't have the willpower to do a dissertation. #truth
>
>

Outraged at his fat shaming, the Twitterverse called for his head. The professor tried to backpedal, saying that the post was just part of a "research project." This claim was debunked a few months later by his university when they formally censured him—and barred him from serving on any graduate admissions committee ever again. He ultimately issued this apology: "My sincere apologies to all for that idiotic, impulsive, and badly judged tweet. It does not reflect my true views, values, or standards."[3]

Just because you are a scholar doesn't mean your every tweet will be well thought out. And just because you have the respect of your community doesn't mean you can't lose it.

☞ The takeaway: Being smart can't save you from saying something stupid.

ADDICTED TO SHAME

Some individuals can't seem to stop landing in a heap of trouble. Who can forget the compulsive sexting scandals that destroyed the life and career of former U.S. representative Anthony Weiner? After his first risqué sext was exposed, Weiner was forced to resign his seat in Congress. A second offense was disclosed during his failed 2013 bid for the New York City mayor's office. The final straw (for his wife, at least) came when he sexted a photo while lying next to his young son, leaving his marriage in limbo. But an inappropriate photo allegedly sent to a fifteen-year-old girl may be his ultimate downfall, landing him in criminal trouble.[4] What might have happened if Weiner had been content with an old-school stack of *Penthouse* magazines instead?

"Britney," a twenty-seven-year-old Texan with long, honey-blond hair and large, doe-like blue eyes was relieved. She'd finally found a job to support herself and her daughter—even if it wasn't exactly her dream job. She was to start work in a local day care center, a line of work not known for being high paying or low stress. But as a single mom, she needed the paycheck.

Unfortunately, she took the time to voice her thoughts on social media, assuming that the post would only go out to those who had her best interests at heart. "I start my new job today," she wrote on her Facebook page in April 2015, "but I absolutely hate working at day cares."

When friends expressed sympathy, she added, "Lol it's all good, I just really hate being around a lot of kids."

Sigh.

What happened next was—*maybe*—for the best.

The post somehow spread to a local online yard-sale mom's group, which was offended by her attitude. Strangers who didn't know Britney started a digital cyberattack, calling her a "dumb bitch," saying she had "Bubonic plague," and worse.

"I actually cried," Britney said. "It really hurt because I wasn't trying to offend anybody." Of course, the brouhaha ultimately went viral, and she found herself on the local news, struggling to explain her comments to the world. "It really was a big mistake," she told the local CBS affiliate. "I don't hate children. I have my own… I love her. I was just venting."[5]

When the day care center was alerted to her post, Britney was told not to bother coming in. "I'm so sad," she wrote afterward. "I feel like a failure here, looking at my daughter crying because I'm afraid that I'm not going to be able to find a job because of my own stupidity."[6]

Now, none of us *really* believe that Britney hates children. Most likely, she's a mom doing her best to pay the bills and raise her daughter. But she made one stupid digital decision that will affect her for a very long time.

And one that you, too, could easily make.

Let's face it—most of us don't know, in an off-line context, every single person in our virtual social media circles. You never know when one person will share something with another—then with another, and so on. We can faithfully check and double check our privacy settings (and I hope you do), but, please, never replace your common sense with technology. The sad truth is, even the strictest privacy settings aren't enough to prevent one offended acquaintance from sending a thoughtless comment hurtling into the world. When you share something, you're potentially opening up your post to your friends' friends and beyond.

Has Britney learned a hard lesson? Definitely. "I'm not going to post anything like it ever," she wrote afterward. "No matter how I feel." And today, if you visit her Facebook page, under her current job, it brusquely states: "None Ya Damn Business."

☞ The takeaway: Never assume you're just "among friends."

———

We've all heard the infamous tale of Justine Sacco, a New York City–based publicist who, as she was about to take off on an airplane to South Africa in December 2013, tweeted the following message to her 170 followers:

> Going to Africa. Hope I don't get AIDS. Just kidding. I'm white!
>
> ↩   ⇄   ♥   •••

But do you know the rest of the story?

Taking the comment at face value, the Internet reacted with vitriol. By the time Sacco deplaned eleven hours later, she had been fired from her job at InterActiveCorp, a Twitter hashtag called #HasJustineLandedYet was exploding, and millions believed that she was a bubble-headed "racist bitch." The following day, she would tweet: "I'm seriously having panic attacks wondering if someone is going to harm one of my family members. My life is ruined already. Are you happy?"

Sam Biddle, the *Valleywag* reporter credited with setting off the frenzy by initially reposting her tweet, later wrote, "The minimal post set off a forty-eight-hour paroxysm of fury, an eruption of Internet vindictiveness... Not knowing anything about her, I had taken [the post's] cluelessness at face value, and hundreds of thousands of people had done the same—instantly hating her because it's easy and thrilling to hate a stranger online."[7]

Yet a 2015 *New York Times* magazine piece that delved into Sacco's situation revealed the opposite intent.[8] Sacco, who had family in South Africa, was actually trying (albeit ineptly) to make a satirical comment on America's sense of immunity from the African AIDS epidemic. "Living in America puts us in a bit of a bubble when it comes to what is going on in the third world," she wrote. "I was making fun of that bubble." Sacco says she never imagined her words would get taken at face value.

As Biddle himself later learned, when one of his own tweets set off an inadvertent Internet firestorm, it's all too easy to have what you write be completely misinterpreted by strangers.

"Jokes are complicated, context is hard," Biddle wrote. "Rage is easy."

This is a tricky one, because we generally feel that we are expressing ourselves clearly. But is it possible that what you are about to post could be taken the wrong way? Is it humor that won't translate well out of context? Consider before you click whether your "joke" could be misconstrued. The same goes for emojis. People who are not digital natives may not be familiar with what these symbols mean. In fact, Ellen DeGeneres created a segment on her TV show about this confusion, called "Know the Emoji." What's humorous to one can be offensive to another.

☞ The takeaway: Never assume that your words won't get twisted.

———

Unhappy New Year. That's how one now-infamous Indiana hairdresser felt after her notorious New Year's dinner at Kilroy's Bar N' Grill in downtown Indianapolis. She was insulted when a dispute with her server over her party's $700 bill was brushed aside because someone needed urgent medical assistance. To her, the woman lying on the floor of the bar appeared to be just some "junkie" who'd "overdosed." She headed home, and at 1:51 a.m., took to the restaurant's Facebook page to post this scathing review:

I will never go back to this location for New Year's Eve!!! After the way we were treated when we spent $700+ and having our meal ruined by watching a dead person being wheeled out from an overdose, my night has been ruined!!! Every year we have come to Kilroy's to enjoy New Year's Eve and tonight we were screamed at and had the manager walk away from us while [we] were trying to figure out our bill being messed up. The manager also told us someone dying was more important than us being there, making us feel like our business didn't matter, but I guess allowing a Junkie in the building to overdose on your property is more important than paying customers who are spending a lot of money!! Our waitress when we were trying to ask about our bill being messed up also said, "What do you want me to do, pay your bill for you?" What a great way to talk to a paying customer![9]

As most know by now, the hairdresser was dead wrong.

Restaurant owner Chris Burton swiftly posted his own response, winning the approval of patrons and fellow restaurateurs nationwide. He pointed out that the "junkie" was actually an elderly woman having a heart attack (thankfully, she seems to have survived) and called the hairdresser "cold-hearted and nasty" for her self-centered outburst.

Although she later tried claiming that the post wasn't her own, the damage was already done: the full wrath of the blogosphere turned on the hairdresser. Her employer, Serenity Salon, first tried to publicly distance itself from her comments, then fired her. By the end, even Burton himself felt badly for his opponent. "You know, you've got to be careful what you post online," he told the *Indianapolis Star*. "It can come back and bite you."[10]

Let's keep our emotions, especially those of rage, to ourselves, shall we? If you are having a bad day or are frustrated with a poor customer-service experience, consider waiting until you cool off. The world is not going to fall apart if you wait twenty-four hours. By the cold, hard (and sober) light of morning, you may feel differently. If you still decide to post your critiques to public media, at least try to express yourself in the most diplomatic way possible.

And do try to get your facts straight.

☞ The takeaway: Never put a temporary emotion on the permanent Internet.

———

A Harvard business professor was another aggrieved diner who ran afoul of online etiquette. He'd ordered a meal from his favorite Chinese restaurant in Brookline, Massachusetts. But upon inspecting his take-out receipt, he discovered that he'd been charged a dollar more per item than what was listed on the online menu, for a total overcharge of four dollars. The Harvard prof must have become enraged by what he saw as false advertising, something he addressed in his consulting practice—although typically against much larger industries. He shot off an email to the restaurant, threatening legal action and demanding a refund for the amount due to him under state law—three times the overcharge amount, or twelve dollars.

The restaurant owner exchanged several contrite emails with the professor, first apologizing for having an outdated online menu, then offering to refund the money. Eventually, rebuffed, the media-savvy restaurateur went public, leaking the emails to Boston.com. The online world exploded, never failing to delight in toppling the hubris of a Harvard elite, especially a triple threat who had attended Harvard as an undergrad and a law student, and now worked as a B-school professor.

Online commenters called him "miserly," "a tool," "a loser," and an "a-hole." Although he quickly backpedaled, posting an apology the next day on his website, the story lingers on in cyberspace to this day.[11] Was the damage to his Ivy League reputation really worth the four dollars?

As politicos from both sides of the aisle—from Colin Powell to Democratic operatives—now know all too well, private emails do not always stay private. It may seem paranoid to some, but many professionals are now working under the assumption that anything put in an email could someday be revealed. To have an expectation of discretion from someone you are battling against is especially unwise. Do you really want to shoot off an angry email to your foes, putting heated words into writing? Many people are now taking screenshots of every embarrassing tirade, from texts to Snapchats. This Harvard professor may have been better off going old school, with a phone call politely expressing his views, rather than putting his livelihood in jeopardy.

☞ The takeaway: Write as if the world is watching.

For Halloween 2013, one twenty-two-year-old from Michigan had a rather twisted inspiration—to dress up as a victim of that year's Boston Marathon bombing, wearing a blue running shift, a race number on her chest, and fake blood dripping from her legs. She ran the insensitive idea past her Michigan-area friends, wore the costume to work, and proudly posted images of herself on Instagram. #TooSoon. Twitter users were not amused and hundreds began tweeting at her handle—@SomeSkankinMI.

People at the Boston Marathon died in terror and agony...and you looked at the images and thought "lol funny costume idea"?

One actual victim of the marathon bombing saw the photo and tweeted this:

You should be ashamed, my mother lost both her legs and I almost died in the marathon. You need a filter.

Outraged vigilantes began circulating lifted images of her driver's license, harassing her parents and even making death threats. "I've had voicemails where they want to slit my throat and they want to hang me and tear off my face," she told a BuzzFeed reporter. "I'm just like, I don't even know how to respond to this right now." She also admitted that she was fired from her job.

What may seem humorous among your small circle of local friends may not play among an audience miles away. Be sensitive to current events. Maybe her coworkers found this dark humor hysterical, but did she really think that actual Bostonians would too, only months after a child, eight-year-old Martin Richard, died in that bombing? When the hapless Halloweener was reached by BuzzFeed, she pleaded for forgiveness. "My costume was not meant to disrespect anyone, ever. I am truly sorry to anyone that I may have offended or hurt with this. I know my apology doesn't ever fix anything that has been done, but at least know that I am being sincere. I can't undo my actions or make up for them, but my apology is a start."[12]

☞ The takeaway: Your posts are far-reaching. Think beyond your local borders.

Around the time of the protests following the shooting of Michael Brown in Ferguson, Missouri, one twenty-year-old criminal-justice student in New York discovered that several racist slurs had appeared on her Facebook page. The trouble was, she hadn't posted them. Still, they had been reposted to a Tumblr page called "Racists Getting Fired," devoted to exposing alleged racists to their employers, and the rest of the world. The AMC movie theater where she worked started getting calls from people branding her a racist and demanding that she be fired.[13]*

Eventually, the true story was revealed: her page had been hacked by her ex-boyfriend, a Miami man who wrote the slurs to set her up, most likely by logging into her account. "I said none of those horrible words of hatred and racism," she hurriedly wrote on her Facebook page. "Anyone who knows me, knows I would never in my entire life say anything like that."[14]

Sadly, it's not just virtual strangers who can come along and humiliate you. It can also be someone who knows you better than anyone else. Friends today may be foes tomorrow. It's better to never share your passwords with others, no matter how much you want to trust them. Even your boyfriend? I'd say, *especially* your boyfriend. As many women have learned with revenge porn, betrayal by a former partner is not uncommon. Also, don't leave a public computer logged in to a social media site, giving access to someone unsavory looking to make comments via your profile. Have you ever left your phone unattended? Near a friend

---

* The *Washington Post* reported in December 2014 that "Racists Getting Fired gained nearly forty thousand followers in a matter of days, with fifteen thousand submissions in the first eight hours of the blog's existence."

who has your passcode and a twisted sense of humor? That's an oops moment you don't want to have. Lock 'em up, your phone and your other gadgets. Those few free minutes occupied by someone else's keystrokes could create years of scrubbing your digital landscape.

☞ The takeaway: Never share your passwords, and change them often.

———

A New Jersey high school student was a highly rated football prospect, ranked the state's third-best player and headed to the University of Michigan that fall on a football scholarship. Until he posted on Twitter.

> Imma marry me a bad ass white women someday!
>
> ← ⇄ ♥ ...

> Yo my pops just drove by and splash some jewish lady with mad water...
>
> ← ⇄ ♥ ...

> Says she aint a hoe but she far from a virgin...
>
> ← ⇄ ♥ ...

These and other, more explicit, messages cost him his football scholarship and also got him expelled from his Catholic high school. "This is a Catholic school; things like that cannot happen," his coach told ESPNNewYork.com.[15] "It was totally inappropriate." The player would go on to play for the University of Colorado, and, so far at least, has made no further social media blunders.[16]

Current and aspiring college athletes may not want to accept that schools will be checking their social media, but it is a reality. A Duke University football coach, Derek Jones, recently confirmed that on Twitter: "Prospects should treat their social media pages like a job résumé because we do. It's our #1 source of character evaluation."[17] Some coaches go even further, spying on their own players. Texas Tech coach Kliff Kingsbury admitted on a podcast that his team's coaching staff has gone so far as to set up fake social media accounts, portraying themselves as "cute girls." Once their players eagerly accept the friendship, the coaching staff will monitor what their athletes put out there "behind" Coach's back.[18] Is this catfishing behavior going a step too far? Maybe, but it's something college athletes should be aware of. This tweet from Football University says it best: "Don't let a 140 character tweet jeopardize a $140,000 scholarship."

☞ The takeaway: Your athletic prowess won't save you.

———

The bouquet of roses had barely lost its bloom for the Texas beauty queen who was crowned Miss Teen USA in 2016. Within hours, social media users pounced on the eighteen-year-old for racial slurs she'd made back in 2013 and 2014 on her Twitter account, using the N-word while seemingly jesting with friends.

Though she kept her crown, she was forced to issue an apology on Twitter and Instagram, claiming, "Several years ago, I had many personal struggles and found myself in a place that is not representative of who I am as a person. I admit that I have used language publicly in the past which I am not proud of and that there is no excuse for... I am today a better person."[19] On *Good Morning America* later that month, she explained that she "was being a follower" when she posted those tweets. "I was trying to fit in with my friends, and the word was thrown around in the music I listened to," said the teen queen. "I had no guidance."[20]

We must all post knowing that social media is a permanent record, and though we're bound to make mistakes, we have to take accountability for our actions. You never get a second chance to make a first impression. Today, social media is the digital résumé that people will see first and judge you on.

☞ The takeaway: Never assume that your posts won't come back to haunt you.

———

Imagine the embarrassment of one Bible-quoting, conservative Virginian running for Congress who took a screen grab of his computer and posted it to his Facebook campaign page—only to inadvertently reveal that his browser tabs contained earlier searches for online porn. Why else would there be tabs reading LAYLA RIVERA TIGHT and IVONE SEXY AMATEUR—references to a professional porn star and an amateur one. The "notorious post," as he later referred to it, remained up for hours after the story went viral, reaching two hundred thousand viewers, while the aspiring politico scrambled for an explanation, only digging himself in deeper with outlandish claims of investigating a conspiracy theory. Finally, he just threw an apologetic Bible verse at his followers, thanking them for their support.[21]

Understanding online basics is a must nowadays. If you're unfamiliar with how screenshots, cookies, and passwords work, you really need to brush up your know-how to avoid such a

blunder. A simple apology and deleting the offending photo
might also have served this man well.

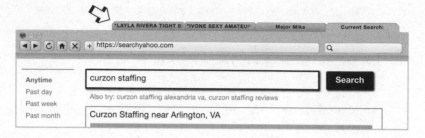

☞ The takeaway: Master your technical skills.

Maybe you've read all the examples in this chapter and are now
feeling a bit smug. *You'd* never post something so dumb, rude,
offensive, or clueless.

Let's try a little thought experiment.

Imagine you're at the gym after a long workout, just getting
out of the shower. Your aging body has seen its share of wear
and tear, but you're not ready to give up fighting that good fight.
You're still rocking it because age is just a state of mind, right?
Plus, it beats the alternative. Sure, your particular gym is filled
with wannabe starlets, models, and reality stars, which would
make anyone self-conscious about those lumps and bumps,
stretch marks, and underarm flab. But, really, that's just your own
insecurities talking. No one else is taking the time to notice. Who
else could possibly care?

Then along comes Dani Mathers, a twenty-nine-year-old

Playboy model who makes her living off her body. Dani Mathers, the 2015 Playmate of the Year, who is apparently incapable of projecting herself into the future and imagining that she, too, might one day happen to age. Dani Mathers, who seems to have all the time in the world to contemplate the deficits of your post-workout body, because she comes along and snaps a naked photo of you for a laugh.

She groans at how the sight of your exposed flesh offends her tender sensibilities.

She writes: "If I can't unsee this then you can't either."[22]

Except she slips. Thinking—or so she later said—that she was sending the image only to a friend (like that makes it better), she instead posts the photo publicly on her Snapchat Story. The news reverberates among her hundreds of thousands of followers on Instagram and Twitter.

Naturally, the Internet explodes, blasting Mathers for her callousness, her callowness, her cruel violation of privacy. Despite her hasty apology that "body-shaming is wrong," she is banned from the gym, loses her radio gig, and is criminally charged with invasion of privacy.[23]

As this sordid tale shows, anonymity when we're out and about in the world is a thing of the past. Step outside your bedroom door, anger or even just amuse the wrong person, and you, too, risk being the victim of a digital disaster.

The backlash against Mathers may be little solace to the woman whose naked image is now forever embedded in thousands of websites. But it goes to show, again, how quickly, how easily, with one little mistake, shamers can find themselves on the receiving end.

☞ The ultimate takeaway: Never think you're safe from digital disgrace.

It Only Takes One Click to Change Your Life

No one is truly immune from making a digital blunder, whether you're an educated college professor or a savvy public relations professional. When putting your words out into the world, you can no longer simply post now and think later. Thoughts can be misconstrued. Privacy can be breached. Friends today can be foes tomorrow. Lovers can turn into boors.

The potential for widespread condemnation, while a relatively new phenomenon for most of us, has been an enduring and pervasive element of celebrity culture for years. While we're only just getting a taste of it, high-profile figures have always dealt with public scrutiny—and now that we can all join in online, the volume—and the content—is even more intense.

CHAPTER 4

SHAMING, CELEBRITY-STYLE

"Where is it coming from that it seems okay to sit in your room and cause pain to somebody, for the sport of it? When did we become that?"

Renée Zellweger[1]

Celebrities—males and especially females—are perhaps the biggest targets for online shaming. Generations of stars have endured the sniping and scrutiny of gossip rags, from that of Louella Parsons (America's first Hollywood gossip columnist) to modern versions like PerezHilton.com and TMZ. But today's online epidemic of hate has left them directly exposed to their millions of fans—or rather, their antifans, whose comments can be fanatically brutal.

Women in Hollywood are routinely criticized for looking too

old—or turning to extreme plastic surgery to stay young. They are shamed for being too heavy (think Melissa McCarthy and Rebel Wilson), for being too thin (like Tara Reid and Keira Knightley), or even for having, allegedly, recently eaten a hamburger (like Selena Gomez, Kelly Clarkson, and Anne Hathaway). CNN's website has featured a running slideshow tallying celebrities who have been body shamed, which as of this writing stands at twenty-nine and counting.[2] Even across the pond, a British TV show dedicated to the genre, called *Celeb Trolls: We're Coming to Get You*, hunts down those who harass celebrities online.

Although some would dismiss this bashing as the price of fame, many stars have revealed that these words do have the power to wound. "I've never been more verbally abused in my life than on Twitter, and specifically in the last few months, having come on this show," observed actress Candace Cameron Bure, after she joined *The View* during the 2015 reboot of her sitcom *Full House.* "You don't have to verbally abuse me and rape me. That's what they do to me on Twitter."[3]

When Twitter comments were made about how the late Carrie Fisher's body had aged between the original and current Star Wars films, she responded fiercely, revealing how she felt.[4]

Carrie Fisher @carrieffisher

Please stop debating about whether or not I aged well. Unfortunately it hurts all 3 of my feelings. My BODY hasn't aged as well as I have.

Michael Jackson's daughter Paris vented her feelings in a video she posted to Instagram, blaming a past suicide attempt partially on Internet trolls. "I've tried sticking up for myself," she said. "I've tried the whole 'blocking the haters' thing, not reading the comments...ignoring it. But it's hard...when there's so much of it."[5]

Even celebrity babies are not immune from coming under attack. After Beyoncé and her four-year old daughter, Blue Ivy, walked the red carpet at the 2016 MTV Video Music Awards, Twitter users trashed the child's looks:[6]

> So are we all just supposed to pretend that Blue Ivy isn't ugly as hell forever?

> Blue Ivy is ugly as sin and there's no way around it.

> I'm sorry but I think blue ivy is an ugly baby.

"Stars have to develop a thick skin," says publicist Howard Bragman, founder of the Hollywood PR machine Fifteen Minutes.[7] "We live in polarized times and everybody's got an opinion." How does he help his celebrity clients get through the dreck? "I try to keep them away, not let them read the crap. It's writing on the bathroom wall."

Those who once found social media a welcome place to connect with fans are increasingly turning sour on the experience. Singer Carrie Underwood has become accustomed to online trolls, who wrote "fake" and "drag queen," beneath a selfie she posted (and later deleted) on Instagram, and tweeted that her look at the Country Music Awards was "satanic."[8] She told *Redbook*, "I used to feel like I could go through social media and talk to people, really have that communication. But you get to a point where there are too many mean people saying mean things—probably just to get a reaction from you—and eventually I was like, 'I don't know if I can do this.' You have to have a barrier up, which is sad."[9]

During the 2016 holiday season, talk show host Wendy Williams shared a Christmastime "throwback" photo of herself on Instagram as an "awkward twelve-year-old," and was likely disheartened to find herself slammed and the subject of mean memes.[10]

Some celebrities have taken to protecting themselves by outsourcing their social media accounts to their underlings. "It truly wasn't a safe space for me," *Girls* creator Lena Dunham said in a podcast interview, reporting that she would no longer personally look at her Twitter account but have assistants handle that unpleasant task. "I think even if you think you can separate yourself from the kind of verbal violence that's being directed at you, that it creates some really kind of cancerous stuff inside you."[11]

Just as we hear about high-profile writers and bloggers quitting social media, reporting on the latest celebrity "flounce" has become a daily ritual. Fifth Harmony singer Normani

Kordei vocally quit Twitter after she received racially tinged harassment from her band's own fans, who were unhappy with a somewhat lukewarm endorsement she made of a fellow band member. "I've not just been cyberbullied, I've been racially cyberbullied with tweets and pictures so horrific and racially charged that I can't subject myself any longer to the hate," she wrote.[12] But, like us, celebrities often can't keep away. The primal desire to promote and connect with fans ensures that most end up returning for more. A month later, Kordei was back, now serving as a diversity ambassador for the antibullying nonprofit Cybersmile.

Justin Bieber also deactivated—and then reactivated[13]—his Instagram account of 77 million followers (as of this writing, the platform's sixth-ranked account) for two weeks in August 2016. What ticked him off? Bieber was seemingly irritated after photos of his seventeen-year-old new flame sparked negative reaction from fans and an online exchange of words with his ex, Selena Gomez.[14]

Some celebrities don't join at all, like Jennifer Aniston. But that doesn't mean she isn't part of the social media landscape. On the contrary, she told the *Daily Express* that cyberbullying has turned her into a social hermit. "You do the best you can, but it feels like it's getting worse and very nasty because of bullying on the Internet," she said. "Even movie critics don't just comment on the film, they tear people apart on a human level. I don't know why that is happening but it is... We just stay home so there really isn't anything to report."[15]

Star Power Fights Back

Of course, for all their elaborate hand-wringing, celebrities also have options that most of us don't. They can turn to traditional media outlets to address critics or set the record straight. When former *E! News* anchor Giuliana Rancic was widely criticized for looking anorexic, she told *People* magazine that her cancer-suppressing medication was the true cause of her thin frame.[16] The media also came in handy when she was attacked online for what many believed was a racist insult on the show *Fashion Police* about Disney star Zendaya's dreadlocked red-carpet look. After using social media to voice her apologies for a major misunderstanding, Rancic went on the *Today* show to explain that editing had cut out her true meaning, a reference to hippie culture and the singer's bohemian-chic look.[17]

Celebrities can also make direct pleas to their fans for compassion. When '80s child star Corey Feldman debuted his song "Go 4 It" on the *Today* show in September 2016, viewers savaged Feldman's disjointed dance moves, his odd black hood, and his singing style. He responded with a tearful Facebook Live video. "We put ourselves out there and we did the best that we could. And, like, I've never had such mean things said about me… These things that are said about us are awful… I'm sorry if it's not good enough for you, but you don't have to beat us up… Why is it okay to, like, publicly shame us?… It's not okay, it's not acceptable to call us freaks, weirdos, losers, whatever."[18]

Feldman was invited back and made a second appearance on the show days later, defying his critics.

"I don't think [celebrities] have an obligation [to respond to attacks], I think they have an opportunity," says Bragman. "Whether they choose to respond is totally their business. It's good to be smart in your response. Whatever your response, it's good to think about."[19]

But are these public spats genuine? To be sure, if a celebrity chooses to respond publicly, it is often the decision of his or her entire public relations team. As Amanda Hess wrote in a *New York Times* piece on celebrity cyberbullying, "What looks like a public display of immaturity can actually be part of a sophisticated image management strategy. Retweet counts and Instagram followers are the new Billboard 100, and celebrities can gin up their numbers by instigating feuds with one another in increasingly nasty or technologically intriguing ways."[20] That's not to say that, occasionally, celebrities don't go rogue. "Sometimes they don't tell the publicist, they just do it themselves," admits New York City–based publicist Heidi Krupp, CEO of Krupp Kommunications. "Then the team has to clean it up, or delete it, or reposition it to rectify it. That can be a big nightmare."[21]

After a hiatus from the limelight, Renée Zellweger's rare red carpet appearance in 2014 at *Elle*'s Women in Hollywood Awards lit up the blogosphere. Observers couldn't get over how her face seemed so different from her Bridget Jones days. News organizations like CNN felt compelled to post an online poll, asking viewers what they thought of her "new look" (63 percent said they "didn't recognize her"), and Gawker ran photos of her, past and present, to further emphasize the point.[22] Commenters on Twitter

called her "a completely different person," gasping at the fact that
her face dared to change over the course of a decade.[23]

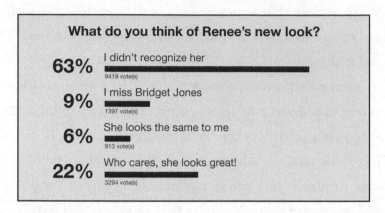

On the *Today* show, Zellweger claimed to be untouched by
the unkind words. "I don't do any kind of social media, so I don't
see it," she said.[24]

But do we really believe this? The public debate reignited
in 2016, when her new film, *Bridget Jones's Baby*, was slated for
release. *Variety* critic Owen Gleiberman mused about what
drives celebrity women to undergo the knife in a piece titled
"Renée Zellweger: If She No Longer Looks Like Herself, Has
She Become a Different Actress?"[25] In a guest column for the
Hollywood Reporter, fellow actress Rose McGowan came to her
defense and publicly took Gleiberman to task. "What you are
doing is vile, damaging, stupid, and cruel," she wrote.[26]

Finally, we heard from a frustrated Zellweger herself, in an
eloquent piece for the Huffington Post, denying that she ever
had eye surgery, botched or not. Mostly, she spoke out against the
narrowly defined parameters of beauty, arguing that they send a

message that "undoubtably [*sic*] triggers myriad subsequent issues regarding conformity, prejudice, equality, self-acceptance, bullying, and health… Maybe we could talk more about why we seem to collectively share an appetite for witnessing people diminished and humiliated with attacks on appearance and character," she wrote.[27]

Were the comments about Zellweger's face out of line? Definitely.

But was her savvy response also well-timed, with three upcoming films—her first in six years—soon to be released? Possibly. As Bragman says, "Controversy doesn't necessarily hurt careers; I've seen a lot of times when it actually helps."[28]

Refusing to Be Shamed

Another downside of being a female celebrity: you're more likely to have your naked photos hacked and posted online, as stars from sportscaster Erin Andrews to actress Scarlett Johansson have discovered. In 2014, the infamous news broke that a ring of hackers was trading noods of female celebrities, including Kate Upton, Kirsten Dunst, and Jennifer Lawrence, stolen off of Apple computers using a "sophisticated phishing scheme." One Pennsylvania man was eventually caught and pled guilty to "unauthorized access to a protected computer to obtain information."[29] The scandal was called "The Fappening," after its subreddit thread name. "For their victims, it is no laughing matter," Edward Lucas wrote of the scandal in his book *Cyberphobia*.[30] "Even the most energetic and expensive legal response cannot scrub the stolen photos from the Internet. As fast as you persuade or order one site to take them down, another puts them up. You can never be sure that they will not appear again—someone, somewhere, has them on his computer, and publishing them takes just a couple of mouse clicks."

In a *Vanity Fair* interview, a defiant Jennifer Lawrence refused to carry the mantle of shame for taking the nude photos. "I don't have anything to say I'm sorry for," she said. "I was in a loving, healthy, great relationship for four years. It was long distance, and either your boyfriend is going to look at porn or he's going to look at you."[31] She blamed the hackers, as well as those who looked at the photos, for what she rightly called a sex crime. "I can't even describe to anybody what it feels like to have my naked body shoot across the world like a news flash against my will," she said. "It just makes me feel like a piece of meat that's being passed around for a profit… Anybody who looked at those pictures, you're perpetuating a sexual offense. You should cower with shame."

Technical Assistance

Stars do seem to get the velvet-rope treatment in one area— when they bring abusive behavior to the attention of social media and tech companies. In the summer of 2016, then Breitbart editor Milo Yiannopoulos, who has been called the "most famous troll in the world,"[32] posted a scathing review of the girl-power reboot of *Ghostbusters*. His critique kicked off a brutally racist Twitter campaign, in which his followers taunted African American actress Leslie Jones and called her a gorilla.

Jones began retweeting her online harassers, in an attempt to report and expose them.

Soon, she received something most of us would not—a personal message from Twitter's cofounder and CEO, Jack Dorsey.[33]

Leslie Jones @Lesdoggg

Twitter I understand you got free speech I get it. But there has to be some guidelines when you let spread like that.

jack @jack

@Lesdoggg Hi Leslie, following, please DM me when you have a moment.

Supporters started the hashtag #LoveForLeslieJ, but it was too late. "I was in my apartment by myself, and I felt trapped," Jones told *Time* magazine. "When you're reading all these gay and racial slurs, it was like, I can't fight y'all. I didn't know what to do. Do you call the police? Then they got my email, and they started sending me threats that they were going to cut off my head and stuff they do to 'N-words.' It's not done to express an opinion, it's done to scare you."[34]

Eventually, Jones signed off from Twitter. "I leave Twitter tonight with tears and a very sad heart," she tweeted on July 19.

But not forever.

The media company was finally roused to action, issuing a statement that referenced Jones's reports and taking the unusually extreme action of permanently banning Yiannopoulos from Twitter.[35] Jones would tentatively return to the site, but sadly, the fight was far from over. In a brutal retaliation, Jones's personal website, justleslie.com, was hacked, with scans of her passport, California driver's license, and nude photos all posted.

As painful and potentially embarrassing as this was for Jones, she flexed her celebrity muscles to have the last word (for now). "I don't know if you all know this, but I ain't shy," she told *Saturday Night Live*'s Colin Jost during a "Weekend Update" segment. "If you want to see Leslie Jones naked, just ask. Just ask!"[36]

Filing Suit

Female sportscaster Erin Andrews suffered one of the most embarrassing examples of public shaming. The former ESPN star's personal space was violated in 2008 when a man illegally rigged a peephole in her hotel room and filmed her changing. The nude video he acquired and later posted online in July 2009 went viral, with millions of views worldwide. Her stalker, Michael David Barrett, was eventually arrested by the FBI and sentenced to two and a half years in prison.

Andrews went on to file a lawsuit against her harasser, as well as the hotel owner and operator, who she argued had failed to protect her privacy by revealing what room she was in. In March 2016, the case went to the jury and Andrews, now a Fox

Sports reporter and cohost of *Dancing with the Stars*, testified as to how deeply she had been affected. "This happens every day of my life," she said. "Either I get a tweet or somebody makes a comment in the paper or somebody sends me a still video to my Twitter or someone screams it at me in the stands and I'm right back to this. I feel so embarrassed and I am so ashamed."[37] Andrews would ultimately be awarded $55 million for her pain and suffering.

Celebrity activist and actress Ashley Judd, known for being a huge Kentucky Wildcats basketball fan, found herself under attack after she posted a tweet about a rival sports team's "playing dirty" during the March Madness tournaments.

As a high-profile actress not afraid to speak her mind, Judd has long been the victim of a barrage of abuse that many female journalists, feminist writers, gamers, and sports reporters now sadly consider commonplace in today's online environment. "I have responded to this with various strategies," she said in a January 2017 TED Talk. "I've tried engaging people… I've tried to rise above it, I've tried to get in the trenches, but mostly I would scroll through these social media platforms with one eye partially closed."[38]

Judd does acknowledge that as a celebrity, she receives special treatment. "I have all these resources that I'm keenly aware so many people in the world do not… I can often get a social media company's attention… I actually pay someone to scrub my social media feeds, attempting to spare my brain the daily iterations of the trauma of hate speech."[39] Judd, a rape and sexual abuse survivor, contends that her critics often claim she just needs to grow thicker skin. That she should accept this because "I'm famous. It's part of my job description."[40]

Still, what happened that March went well beyond the norm. "When I express a stout opinion during #MarchMadness I am called a whore, c—, threatened with sexual violence. Not okay," Judd tweeted on March 15, 2015. "The volume of hatred that exploded at me in response was staggering," she later wrote. "Tweets rolled in, calling me a cunt, a whore or a bitch, or telling me to suck a two-inch dick. Some even threatened rape, or 'anal anal anal'… My age, appearance, and body were attacked."[41]

ashley judd @AshleyJudd
When when I express a stout opinion during #MarchMadness I am called a whore, c---, threatened with sexual violence. Not okay.

Now, did Judd open herself up to harassment with her trash-talking tweet? Some may say maybe. She took the post down. But as one observer noticed, a man engaging in similar sports talk would never face the same level of intensive gender-directed abuse. Judd went on to write a feminist op-ed titled, "Forget Your Team: Your Online Violence Toward Girls and Women Is What Can Kiss My Ass," even against the wishes of her own publicity team. "My chief advisor said, 'Please don't, the rain of retaliatory garbage that is inevitable—I fear for you.' But I trust girls and I trust women, and I trust our allies. It was published, it went viral."[42]

And Judd says she will hold her abusers responsible for their threats.

"Everyone needs to take personal responsibility for what they write and not allowing this misinterpretation and shaming culture on social media to persist," Judd told MSNBC. "And by the way, I'm pressing charges."[43]

We are still waiting to hear news of further action against her attackers. Celebrities like Judd and Andrews, with their extensive resources and high-profile platforms, have perhaps the best chance of taking up this cause for the rest of us.

Hate Shaming

Stars may receive the brunt of online attacks, but they also have the option of using their star power to shame their shamers right back. Comedian and actress Margaret Cho spoke in an *Observer* article about taking a specific action that she dubs "hate shame."

> People often say, "Oh, don't feed the trolls." I screenshot what they say, report it, and I'll send it to their employers, their spouses— things of that nature, where you're calling attention to the hatred and cutting it off at the same time with immediate hate shame. If you hate shame, it automatically forces them to revisit the hatred they're spewing, that can be potentially very damaging. And using it against them—as opposed to internalizing the

trolling and feeling uncomfortable with it—it's a good opportunity to stand up for yourself and also bring reverence to the
kind of racism, misogyny, and homophobia that these people
perpetuate. We should all practice this type of self-defense.[44]

Like Cho, CNN's Anderson Cooper shared on *The View*
that he uses similar tactics when dealing with his abusive trolls,
reviewing their social media profiles and using what he reads to
call them out on their bad behavior. But one of the most poignant
moments of response came when a celebrity dad was compelled to
stick up for his little girl. In August 2016, actor and director Kevin
Smith took to Instagram to call out a person who had insulted
pictures of his seventeen-year-old daughter, Harley Quinn Smith,
by posting "you're ugly as shit" and "you're cancer." Smith wrote:

If you hate me (or my kid) this much, the better use of
your time is to make YOUR dreams come true, instead of
slamming others for doing the same. The best revenge
is living insanely well—so if you wanna get back at a
17-year-old girl for the grievous crime of enjoying her life,
the best way to do it is to succeed in your OWN existence.
Show the world WHY we should be paying attention to you
instead of anyone else.[45]

Inspiring Change

What's important to remember about celebrities is that they have the ability to use their platform of fame to make a difference—and many of them do.

Ariel Winter, the young actress who plays Alex Dunphy on *Modern Family*, has been repeatedly body shamed for her voluptuous figure. When she celebrated her high school graduation with a photo on Instagram, some focused on her full figure and her pink dress, with its revealing cutout. "Who goes out in public like that!!," wrote one commenter. "Gosh I hate to say it but do 18 years olds [*sic*] really dress like this? Lovely dress and body but it doesn't give anything to the imagination."[46]

Winter responded with this tweet:[47]

Ariel Winter @arielwinter1
Dear sorry body-shamers, I looked HOT in that dress. And if you hate it, don't buy it. But please get a hobby. XOXO Ariel #EmbraceYourBody.

And when she was criticized about wearing a bikini—at the beach, no less—Winter responded this way on her Instagram feed.

I typically never give power to the mean things people bravely say behind their computer screens on the Internet, but this is for the girls who are constantly bullied, whether it be online or at school. You are not asking for anything because of what you are wearing—you are expressing yourself and don't you ever think you deserve the negativity as the consequence to what you are wearing—YOU ARE BEAUTIFUL. Celebrate you and don't let anyone's comments allow you to think less of yourself. Us girls have to stick together!!!!!![48]

Winter is now a Dove spokeswoman for the brand's #SpeakBeautifulSquad, speaking out against this form of body shaming.

Singer, songwriter, and actress Lady Gaga was thrilled to be invited to perform at the 2017 Super Bowl Halftime Show. The audience was just as energized, and the star didn't disappoint, opening her show with a patriotic sky of drones as she flew down into the stadium. Social media was buzzing, and many celebrities such as Ellen DeGeneres, Adele, Tyler Oakley, and Chris Pratt—along with many others—were quick to send Gaga their tweets of approval.

Tyler Oakley @tyleroakley
the versatility being showcased right now... this artist is once in a lifetime. we are so blessed to exist at the same time. #PepsiHalftime

Lady Gaga was not only feeling the love from her peers: People from all over were sending out messages of support and admiration. Many were posting they were going to start downloading her music and commenting that her performance was a reminder of her positive impact. As a matter of fact, her music sales spiked one thousand percent following her Super Bowl performance.[49]

With all this cyberadmiration, though, came those who were ready to burst the balloon and deflate the glory. In 2012, Lady Gaga revealed that she is an eating disorder survivor. She admits she works hard to stay in shape; however, with the benefits of her parents' restaurant, her weight does fluctuate.

So what did the Internet trolls have to say about a performance that was globally loved by so many? The online bullies were more concerned about her *belly* than the phenomenal halftime show.

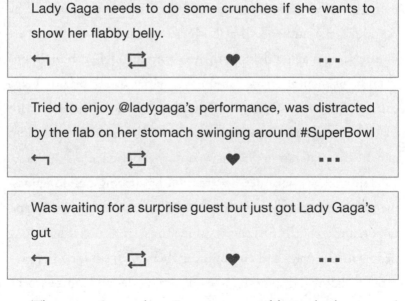

Lady Gaga needs to do some crunches if she wants to show her flabby belly.

Tried to enjoy @ladygaga's performance, was distracted by the flab on her stomach swinging around #SuperBowl

Was waiting for a surprise guest but just got Lady Gaga's gut

The superstar took to Instagram to address the haters and her fans:

I heard my body is a topic of conversation so I wanted to say,
I'm proud of my body and you should be proud of yours too. No
matter who you are or what you do. I could give you a million
reasons why you don't need to cater to anyone or anything
to succeed. Be you, and be relentlessly you. That's the stuff of
champions. Thank you so much everyone for supporting me. I
love you guys. Xoxo, gaga[50]

Lady Gaga is committed to supporting the wellness of
young people and empowering them to create a kinder world.
In 2012 she founded the Born This Way Foundation, and in her
panel discussion with Oprah Winfrey at Harvard University
in February 2012, Gaga discussed the importance of not only
helping the victims, but also reaching out to the bullies. She
reminded the audience that both the victim and the bully are
on the same playing field, both are going through a tremendous
amount of mental turmoil and need help.[51] In 2016 Lady Gaga
expanded her Born This Way Foundation by joining forces with
Hack Harassment. Together they aim to fight online harassment
and create a safer, more inclusive online experience.

Lady Gaga continues her mission of curbing online harassment and bullying based on her three pillars: safety, skills, and
opportunity. With this, the organizations are connecting young
people in safe ways and empowering them with skills and oppor-
tunities that will inspire them to create a kinder and braver world.

Although trolls usually like to pick on vulnerable women, occasionally they go after men as well. Actor Chris Pratt shared with almost ten million Instagram followers, "Well, just because I am a male doesn't mean I'm impervious to your whispers. Body shaming hurts."[52] Similarly, actor Wentworth Miller made headlines in March 2016 after a meme went viral, body shaming the actor about his weight. Two photos were posted side by side: the first, from when he starred on the show *Prison Break*, showed him bare-chested, toned, and tattooed. The second, from 2010, caught him wearing a red T-shirt that barely hid a potbelly. The caption read: WHEN YOU BREAK OUT OF PRISON AND FIND OUT ABOUT McDONALD'S MONOPOLY.[53]

Miller wrote an open letter on Facebook to help educate people on depression and suicide.

> In 2010, semiretired from acting, I was keeping a low profile for a number of reasons.
>
> First and foremost, I was suicidal…
>
> In 2010, at the lowest point in my adult life, I was looking everywhere for relief/comfort/distraction. And I turned to food. It could have been anything. Drugs. Alcohol. Sex. But eating became the one thing I could look forward to.

The group that originally posted the meme on Facebook, The LAD Bible, ended up posting this apology: "Today we want to say we've got this very, very wrong. Mental health is no joke or laughing matter. We certainly didn't want to cause you pain by reminding you of such a low point in your life. Causing distress and upset to innocent or vulnerable people is simply not acceptable."[54]

———

Of course, celebrities can also make serious mistakes that bring condemnation upon them. But while we live in a society quick to judge, we can also be very forgiving. Time and time again, we have torn down our beloved idols, only to welcome them back into our lives—provided that they are contrite, humble, and willing to accept responsibility for what happened. Celebrities like Martha Stewart, Paula Deen, Brian Williams, Tiger Woods, and Michael Phelps may always have a smudge on their reputation, but that doesn't stop them from rebuilding their identities.

But what happens to those who are cybershamed and don't have the support of fame, fortune, and followers to lean on?

CHAPTER 5

THE RAMIFICATIONS OF A DIGITAL DISASTER

"I am not the person I used to be before this ordeal. It left me mentally unstable, physically debilitated, and socially isolated."

—Lena Chen[1]

Every day, the words and images we all choose to share can do more than simply embarrass us—they can affect our entire future. There are those who have been fired from their job for a tweet, surrendered a college scholarship for a post, or lost out on a new employment opportunity because of their online presence.

Jobvite's annual "Recruiter Nation" survey revealed that 92 percent of hiring recruiters now review a candidate's social media profile before making a hiring decision.[2] And what they find there can make or break you. More and more employers admit that they are turning to social networking sites to dig up something negative on potential candidates, and what's being revealed is not pretty. A 2016 CareerBuilder survey reported that nearly half of employers who research job candidates on social media said they'd found content that caused them to not hire the candidate, up from 34 percent in 2012.[3]

According to that survey, here's what hiring managers discovered:

+ Candidate posted provocative or inappropriate photographs or information—46 percent
+ Candidate posted information about them drinking or using drugs—43 percent
+ Candidate had discriminatory comments related to race, gender, religion, etc.—33 percent
+ Candidates bad-mouthed their previous company or fellow employees—31 percent
+ Candidate had poor communication skills—29 percent

"In a competitive job market, recruiters are looking for all the information they can find that might help them make decisions," said Rosemary Haefner, chief human resources officer at CareerBuilder. "Rather than go off the grid, job seekers should make their professional persona visible online and ensure any information that could dissuade prospective employers is made private or removed."[4]

This form of screening, of course, also applies to aspiring college students. In 2017, Kaplan Test Prep reported that 35 percent of the admissions officers it surveyed admitted to viewing college applicants' social media postings.[5] What triggered the deep dive? Usually, interest in a student's special talent, to verify an award, or to delve into prior disciplinary action. As we saw earlier, student athletes are increasingly under the microscope and landing themselves in trouble. Many schools are formally outsourcing this monitoring to software companies like Varsity Monitor, whose website claims that it allows coaching staffs to "evaluate a recruit before a scholarship offer is made, looking at up to four years of social media history."[6]

I Can't Believe They Fired Me!

Once you land that job, you're not in the clear. Did you know that many companies use software to monitor their employees' posts and have implemented social media policies about what you can and can't say online? The law firm Proskauer Rose found in its 2013–14 "Social Media in the Workplace" survey that nearly 80 percent of the companies surveyed had a social media policy in place, up from 60 percent in 2011.[7] Just one inappropriate comment about your workplace, or other topics, can get you fired if you cross your firm's guidelines.

But that doesn't always stop us from needing to rant. One study found that approximately half of all workers posted pictures, videos, or messages involving their employer.[8] Remember the twenty-five-year-old Yelp/Eat24 employee, who wrote an open letter griping to the company's wealthy CEO about her low pay? "I can't afford to buy groceries," she wrote in her widely read post. "Isn't that ironic? Your employee for your food delivery app that you spent $300 million to buy can't afford to buy food."[9] She ended up getting fired within hours, explaining in a later tweet that she had apparently violated her company's "terms of conduct."[10]

Some on-the-job missteps should simply be obvious—like the two Domino's workers in North Carolina who, in 2009, uploaded a YouTube video of themselves flagrantly violating pretty much every health code in existence—stuffing cheese up their noses and spreading snot on the sandwiches. Though it was clearly a goof and the man and woman involved later claimed that no customer was given the food they mishandled, they both lost their jobs and faced felony charges of food tampering, and the storefront was forced to temporarily shut down.[11] The woman, a thirtysomething mother of two, apologized for the stunt, writing an email to the pizza chain that Domino's executives shared publicly in an attempt at damage control. Not surprisingly, she lamented to ABC News a month later that she was having a hard time finding another fast-food restaurant in the area that would hire her. Similarly, in 2016, fast-food workers at a Philadelphia Checkers were fired after they posted a forty-four-minute Facebook Live video boasting that they spit in the food and washed the floor with burger buns. One, who later claimed that it was just a publicity stunt for his music career, tried to delete it, but it was downloaded and reposted anyway.[12]

It's not just hiring managers and the HR department

assessing your online persona. Your coworkers are equally likely to check you out. A June 2016 Pew Research Center survey found that 17 percent of employees go to social media to learn more about a coworker and another 17 percent to strengthen their connection with a coworker. But frequently, that impulse can backfire: 29 percent of young workers discovered something online that lowered their opinion of that colleague.[13] Did you really want your coworkers to find out what you're into?

THE $28,224 TWEET?

Seattle Mariners reserve catcher Steve Clevenger waded into the culture wars, posting his opinions with a series of tweets calling the Black Lives Matter campaign "pathetic." Not smart. "We strongly disagree with the language and tone of his comments," said the Mariners general manager. He was suspended for the rest of the season, forcing him to forfeit ten days of his $516,500 annual salary. According to the Associated Press, that would cost Clevenger $28,224.[14] Freedom of speech may be priceless, but Clevenger paid a high price per character indeed.

Even comments unrelated to work can cause blowback for your job. Remember the slogan, "What happens in Vegas stays in Vegas"? Many people still seem to be operating under the misconception that what they do online stays online. But as some have learned, that is no longer true. Employers today aren't drawing a distinction between their employees' online personas and their real identities. If you have something nasty to say, in our country, you are free to express yourself. But that doesn't protect you from holding on to your job, since employers see their workers as an extension of their own brand.

In November 2015, a hotel chain supervisor commented on Australian newspaper columnist Clementine Ford's public Facebook page, calling her a slut.[15]

Ford, an outspoken feminist writer whose page currently has 150,000 followers, has often been the target of harassing comments from men, whom she has taken to publicly shaming in turn. She called this man out on his comment, pointing out his behavior to his employer, Meriton Group, who investigated the claim and then turned around and publicly fired him.

"These men have rarely ever faced consequences for their actions," she observed afterward, facing down critics of her response, "but that's starting to change."[16] Just one unsavory word was enough to put this man on the unemployment line— possibly for a good long time, now that this story has been reported by Mashable, The Daily Beast, and the *New York Times*, among others.

Other unsavory comments online are no longer being tolerated. One teacher's aide at a Georgia elementary school found her Facebook posts gutted after she allegedly called former first lady Michelle Obama "a gorilla," linking the post to an article dubbing Obama the most admired woman in the world and writing, "I admire a gorilla more than I admire her (Wait, I forgot, she is a gorilla)." Online observers reposted her comments, asking, "Is that who you want teaching your kids?"[17] Soon after, the district reported that she had been relieved of her duties.

In another case, a Florida assistant state's attorney was suspended after he posted an insensitive Facebook rant in the wake of a tragedy. After news broke about the massacre at the Pulse nightclub in June 2016, he described downtown Orlando as "a melting pot of third-world miscreants and ghetto thugs," then added, "All Orlando nightclubs should be permanently closed. With or without random gunmen they are zoos, utter cesspools of debauchery." The DA's office ultimately fired him, claiming that his posts violated its social media policy. Will cases like this hold up? The attorney said he plans to appeal, arguing that the policy "is vague, arbitrary, lends itself to subjective interpretation, and has a chilling effect on the right of my protected free speech."[18]

Ding, Dong, Process Server
Some People Call Them Posts, Attorneys Call Them Exhibit A

Everything you send or post online has the potential to end up an exhibit in court someday. Not everything you say or type online is appropriate or allowed. You have to learn the boundaries of free speech. Defamation is not condoned as part of free speech. Opinions or satire are one thing, but if you're presenting something that is incorrect and damaging to an individual's reputation as if it's a statement of fact, it may land you in legal trouble.

Even a simple click can potentially be incriminating if it's been deemed off limits. After a nasty divorce, one New York woman, "Mariela," was under a restraining order to avoid all forms of contact with her former husband's family, including her former sister-in-law. When she wrote the word "stupid" in a Facebook post and tagged her sister-in-law's name, a judge ruled that it counted as electronic communication, akin to sending an email or making a phone call. For violating the court's protective order, Mariela is now facing up to a year in jail.[19]

What you post could also trigger an investigation that could land you in serious trouble. Life insurance companies are now using social media to screen clients' applications. Did you conceal that you're an avid scuba diver, while your spring break vacation photos give you away? Claim to be a nonsmoker, but your friend tags you in a photo with a cigarette between your lips? File for a workman's comp claim for a back injury, then "check in" on the ski slopes?[20] A charge of insurance fraud could do permanent damage to your online and off-line reputation.

Ill-advised posts can also come back to bite you in family or divorce court. Too many parents involved in custody battles have found that seemingly small things, like a Facebook photo of themselves with a drink in hand or a description of their lifestyle on a dating site like Match.com, have been allowed into evidence as an example of poor character. "A lot of divorce is 'he-said, she-said' stuff," explains Kristin Zurek, a family law litigator with Cordell & Cordell in Missouri. "Judges or child advocates are always looking for other sources of information, so naturally social media posts or platforms are where people go to get that.

Don't put anything in writing you don't want to see handed to you with an Exhibit A sticker." Sound obvious? "One lady was asking for support, saying she was bedridden and too sick to work, but someone posted a picture of a party at a bar—and there she is, dancing on a table and holding a bottle of alcohol," Zurek says. "People just don't think."[21] Another woman's boyfriend nearly cost her custody when his gun-pride Facebook posts were raised as a safety concern.

Zurek tells clients not only to stop posting on social media while their case is active, but also to go back and tighten up their history. Locking down your privacy settings may not be enough— opposing counsel may play dirty and have your ex convince a mutual friend to access your account and screenshot anything damning. She also recommends telling your wider network to be discreet. One of her clients told the court that his new girlfriend wasn't moving into his home—but the truth was exposed when she posted her new address on Facebook and bragged about moving in with him. "Your network needs to know what you're going through," she says, "and to stop posting."

Emotional Ramifications

We have seen some of the potential professional and legal effects of online shaming, but what about the emotional side? While much attention has rightfully been paid to the severe effects of bullying and cybershaming among our youth, the impact on adults can be devastating too. What does it feel like to undergo an online attack?

We already know that traditional bullying can lead to emotional distress: the Workplace Bullying Institute's 2012 survey reported that the stress from bullying is associated with emotional disorders such as anxiety and panic attacks, as well as physical ones, from loss of sleep to stress headaches.[22] Ellen Walser deLara, a family therapist, professor of social work at Syracuse University, and author of the book *Bullying Scars: The Impact on Adult Life and Relationships*, writes that adults bullied as children or teens can suffer for *years afterward* with trust and self-esteem issues, as well as psychiatric problems. She calls the phenomenon adult post-bullying syndrome.[23]

Being harassed online can leave many victims with the same psychological damage—or even worse. "Make no mistake, the pain of cybershaming can wreak havoc on mental health," says Dr. Michele Borba. "Shame can be debilitating. It derails your entire reputation, especially online, you have no idea who knows. It hits the core of your identity, every part unravels, you're ravished. After a while, you start to say, 'Do I deserve this? Did I do wrong?'" And there is no age limit to online angst, Dr. Borba says. One seventy-one-year-old woman described her emotions to her this way: "If I could just talk with someone that has dealt with a situation like this. It is absolutely smothering. She sucked everything except for my breath from my body. My brain, soul, heart [have] been bruised."[24]

"It's absolutely traumatizing," agrees Dr. Robi Ludwig. "It feels like it's never going to go away, like your whole world is caving in, like a scarlet letter. It's easy to stay fixated on the nasty things that might even feel true. It doesn't take human beings much to attack themselves."[25]

Lena Chen was a freshman at Harvard when she started penning the soon-to-be controversial *Sex and the Ivy* blog, writing frankly about her hookups, sexuality, and eating disorder. When an ex-boyfriend posted nude photos of her, the reaction around campus was unforgiving, and she found herself attacked by the college community and beyond. She would eventually write movingly in *Time* magazine about how the whole experience had changed her. "I am not the person I used to be before this ordeal," she wrote. "It left me mentally unstable, physically debilitated, and socially isolated. I still get extremely anxious in particular social situations. Despite the outward facade of a busy and active social life, I am actually distrustful of others and fearful of intimacy. I interpret benign gestures and comments as hostile, make excuses to not go out, and wonder too often what my neighbors think of me."[26]

In fact, many psychologists now liken the experience of cybershaming to the post-traumatic stress disorder of a *soldier coming home from war*. "In some cases, the effects are similar to PTSD that a soldier endures in battle—but the war zone is on our computer, iPad, or cell phone," Dr. Borba explains. "It could be weeks or months later that it comes back to haunt you, the sound of the click, or you look at a screen and feel that shame again. It wreaks havoc on your sense of safety and inner security."[27]

Samantha Silverberg, MA, LPC, a licensed clinician who works with the nonprofit Online SOS, says her most important task is to make victims of online shaming understand that they did not do anything to deserve it. "In the moment it feels so overwhelming," she says. "There's a lot of negative self-talk that comes up when you experience something like this. I challenge them with evidence... Normalizing and validating are so important. People think, 'This only happens to me, and it's something about me.'"[28] It's common to go through the first four stages of grief—denial, anger, bargaining, and depression—before finally getting to acceptance.

Since trust is a huge factor, dating and romantic relationships become especially difficult. Those who have been the victims of revenge porn or a sexting scandal often find it difficult to trust ever again, according to psychology professor Michelle Drouin. "People take these transgressions very seriously," she says. "Having someone betray your trust in one of the most private parts of your life is devastating, no matter [your] age."[29]

The stress of shaming can also have physical effects on the body. ESPN reporter Britt McHenry had a public meltdown after her car was towed from a parking lot in Alexandria, Virginia. While retrieving her vehicle, the on-air personality insulted the clerk at the towing office, an exchange that was caught on security tape and leaked online. McHenry received a one-week suspension from her employer and a barrage of abuse on social media, with threats so intense that she filed a police report, fearful for her safety. "It will forever be something I'm embarrassed about and will regret," she later recounted in a piece for Marie Claire.[30] "I know the posts about me will live forever online." All that stress, she believes, was likely a contributing factor to a vision impairment she developed soon after the incident that also lingers to this day. "I could no longer see clearly; everything was a blur… The doctor says the vision in my eye might never improve."

Cyberbombed

In extreme cases, Internet shaming and online hate can affect all aspects of your life—professionally, emotionally, and romantically. "Joseph," a fifty-one-year-old college professor working abroad, ended up embroiled in a vicious court battle when his ex-wife left him for an older and wealthier man. In order to win custody rights of his three children, he says his ex fabricated false accusations of domestic violence. Though Joseph's children defended him throughout the process and ultimately remained living with him, the case made its way up to that nation's highest court, the equivalent of the U.S. Supreme Court. Because of the high-profile nature of the case, even five years later, when you search his name, court documents citing those slurs pop right up. "Because of the volume of hits on that website, we cannot get this off the front page when people Google me," Joseph said in an interview.[31] "As a result, I have lost jobs, I was unable to get interviews. I find myself in a constant place where I have to explain something that did not happen. It's one of the first questions I get asked."

When seeking employment, Joseph has found himself in a "precarious position," dammed if he does, and dammed if he doesn't. If he brings it up proactively, doors slam in his face. If he goes mum, it looks like he's hiding something. He's considered changing his name, but he's reluctant to do so, because that would wipe out all his publishing credits and work résumé.

On the romance front, his online reputation became an issue when he was back on the dating scene. Even though he ultimately remarried a woman whom he had known for years, lingering doubts have still damaged his relationship with his new wife. Community members will find and anonymously forward her the document, sowing small seeds of distrust between them. "It does affect my marriage," he confesses. "My wife is human; there are moments where she says, 'Tell me again, show me again.' It's horribly frustrating."

For now, Joseph has relocated to a small town. He has started his own firm and doubts that he'll work in higher education again. "It's a virtual reality that I cannot [make] go away," he says. "I feel like I can't take steps forward, because I have this huge banner over my head." He has spent thousands of dollars over the past several years, working with three different reputation management firms and petitioning the foreign court to redact his name or remove the records of his case from its site, all to no avail. "It's tainted me," he says, a little bitterly and a bit resigned. "It's going to be there with me for the rest of my life."

Joseph's experience is extreme—most of us won't have our divorce proceedings wind up in the Supreme Court. But this is happening on a smaller scale every day. "I've had twenty-five people come to me in last couple years [for advice]," Joseph said, "because they saw what happened to me."

Sports journalist Jen Royle knew right from the start that her abrasive personality might attract critics, but she never expected what happened to her. The single, thirtysomething, Emmy-winning journalist had just finished covering the New York Yankees 2009 World Series win and was ready for a new challenge, so she avidly pursued a job reporting for the Mid-Atlantic Sports Network (MASN) in Baltimore. But from the start, her blunt-spoken coverage and newbie mistakes as a rookie covering pro football angered a vocal subset of male Baltimore sports nuts. "The fans hated me," Jen, a blue-eyed, willowy brunette with the mouth of a sailor and tattoos to match, told me when we spoke. "Looking back, I understand why. I can be abrasive, for sure."[32]

That's an understatement. Jen is admirably tough, the kind of woman you don't want to cut off in traffic. She tells stories about making a female reporter cry for spreading sexual innuen-does about her, and the time she questioned the size of one of her Twitter attackers' genitals. But fighting back brings the risk of inciting your audience even further. "Did I come across as a know-it-all? Maybe," she admits, sitting in a café in Boston's North End. "But I did know—not all of it, but a lot. Everything I said about the team, I was right. They just didn't want to hear it. Because I had the balls to tell it how it was."

After a year at MASN, Jen jumped at a new opportunity, hosting the show *Baltimore Baseball Tonight* for the CBS sports radio affiliate, The Fan, but the mob mentality on social media only grew, egged on, she says, by reporters at a competing sports radio station. (She eventually filed a lawsuit alleging defamation, a suit she later dropped.) Soon, there were fake Twitter accounts and hashtags ripping her apart, one called #FuckJenRoyle, another falsely speculating that she had AIDS. "People telling me to kill myself, and tweeting how ugly I am… It was just too much. They told me I had a big nose, I'm fat, and I have no boobs. I was pretty secure with myself, but every once in a while, you go"—she raises a hand and gestures self-consciously to her nose—"Of course you do."

She thought that if she could only meet her critics face-to-face—in other words, break down the online disinhibition effect—she could charm them, or at least shame them into decency. "I said to some of my haters on Twitter, 'Why don't we all go have drinks—I'll pay the bill,'" she recalls. "'I want to get ten of you around a table, and let's see if you'll hate me after.' I thought it was the best experiment. And no one would do it!" Instead, the attacks were relentless, and Royle began crying "twenty hours a day," locking herself in a closet, and essentially "having a nervous breakdown."

After nearly two hard years in Baltimore, she was ready to flee, accepting a job back in Boston and agreeing to be interviewed by the *Baltimore Sun*'s media critic about her departure, on one condition: that the paper disable its online comments section. Things at that point were bad enough that the paper complied. "There is a line that has to be drawn between professional and personal attacks. I've said this several times—I hope to God I made things easier for the female coming in after me. And I will never talk about this again."[33] Jen thought she would be sheltered from online hate back in her hometown. Amazingly, though, a subset of Baltimore fans continued the hate chatter, celebrating her departure as lifting a "curse" on the team's losing streak. The owner of a local restaurant, Jimmy's Famous Seafood, tweeted his customers in January 2013:

Jimmy @JimmysSeafood
#RavensNation - Tweet @ ▓▓▓▓ & let her know what a piece of scum she is & receive 52% off dinner now thru Sun!

Then a second wave of harassment launched. For reasons that still mystify her, a small group of trolls targeted her, creating Twitter accounts that mocked her relentlessly for a six-month period, poking fun at her "orange" skin tone, her tattoos, her employment history, and her mental stability:

The Final Jeopardy Answer is: a completely insane, unemployable, litigious, fellatious reporter. You have 140 characters. Good luck!

You can block me but you are toast. #slut

> Being in the presence of women who gargle with semen
> make me stutter. It's a character flaw.

The one referencing semen included her boss's Twitter handle.

"These men spent every waking hour of their day tweeting me or other people and telling them that I am the most disgusting, vile person to walk the face of the earth," Jen recalls. "Everyone told me, 'Ignore these idiots, don't let them make you cry.' I didn't take anybody's advice."[34] Instead, she replied full on, calling one "SICKO," retweeting the comments, and threatening to find out who the trolls were.

Soon, though, the tweets referenced her basement apartment, leading her to fear that they knew where she lived, and she felt unsafe enough to bring her fears to Boston police, who recommended monitoring the situation but not antagonizing her harassers by pursuing a case.

Then an incredible stroke of luck fell into her lap. One woman, disturbed by the online attacks, privately emailed her, saying that she knew the identities of the men, who had befriended one another via the site Boston Sports Media Watch. One lived in the Boston suburbs. Another was on the Cape. One was out in Nevada.

In true Jen Royle style, armed with their identities, she went on the attack, tweeting them directly: "Hi, Rich, how's your mortgage business doing? How's your wife Carol? And your three kids?" Stripped of their anonymity, the men responded with shock, then fury; after lashing out one last time, like true trolls, they slunk back to their caves. "It all went away," Royle explains, both triumphant and a little mystified. "It just stopped and I moved on."

Today, Jen Royle is long gone from the field of sports journalism. On a whim, she auditioned and landed a spot on ABC's reality TV cooking show *The Taste*, with a home-baked creation of chicken pot pie cupcakes. Yes, she knew she would be putting herself back in the public eye and potentially opening herself up to attacks. "I'm fearless," she says. "No Twitter account is going to stop me from doing something." After making it to the final five, she reinvented herself as a professional chef, stopping briefly for a gig chopping antipasto in Mario Batali's Babbo Pizzeria and publishing a cookbook called *Bullied into Cooking*, to benefit Boston public schools. She now runs a private catering business, Dare to Taste, and credits the cooking show with changing her life. "I really wanted to move on," she says.

Was fighting back the right move? Yes, she says—and no. "It escalates. I was giving people more material. At times, I looked for it—what are they saying about me? Now, I wouldn't Google myself if you paid me a million dollars."

———

As we have seen from these stories, what is said about you online has the power to control your future, from your job prospects to your emotional well-being. While there are no guarantees that something similar won't happen to you, there are certain ways to safeguard yourself. Using the strategies for digital discretion outlined in the next chapter, you might just be able to protect yourself and your loved ones from being the next victim of **shame nation**.

PART TWO

PREVENTING AND SURVIVING AN ONSLAUGHT

CHAPTER 6

DIGITAL WISDOM IS DIGITAL SURVIVAL

"The Internet happened to us so fast, we are still figuring out how to use it to communicate civilly and respectfully."

—Diana Graber, cofounder of Cyberwise
and founder of CyberCivics.com[1]

Hopefully by now you're thinking: *Okay, I'm convinced. How do I make sure I'm never swept up in an Internet takedown?* While no one is totally immune, you do have the power to control many of these situations, if you employ my basic strategies of digital citizenship. You need to understand that you have to conduct your online life as you do your off-line life—with discretion and mindfulness, respect and integrity. As someone who learned the hard way what it's like to have your digital reputation destroyed, I want to share with you what I wish I had known back then. First, I'll review my broad guidelines for forging healthy online social connections, then I'll lay out some practical strategies you can use to avoid stumbling into a bad situation.

Guidelines for Online Sharing:

+ **Sharing much too much:** It's about time we realize that not everything we do in our life needs to be documented online. Many of us have become addicted to document- ing practically every breath we take on social media, from eating a doughnut to taking a train ride. Is it any wonder that *overshare* was *The Chambers Dictionary's* word of the year in 2014? Even digitally savvy teens think people are divulging TMI online. In a 2015 Pew Research Center survey, 88 percent of teen social media users agreed that people share too much of themselves on social media.[2] Everyone needs to understand the importance of social sharing for your platform—versus oversharing for your ego. A 2015 UCLA study revealed that people who overshare on social media are at a higher risk of being cybershamed. This study suggests oversharing of personal information leads bystanders to blame and not feel for the victim.[3]

  Parents guilty of "sharenting," beware. As University of Florida law professor Stacey Steinberg points out, "There is no 'opt-out' link for children, and split-second decisions made by their parents will result in the indelible digital footprints."[4] One teen girl from Austria actually sued her own parents for all those embarrassing childhood photographs they posted on Facebook, from shots of her getting her diaper changed to potty training. "They knew no shame and no limit," the girl lamented.[5] If you really wish to share such private moments, consider cherishing them only among loved ones who you are confident have your best interests at heart.

+ **Sharing inappropriate material:** The Internet is unforgiving. Before texting, tweeting, emailing, posting, or sharing anything, consider how you'd feel if your words or images went viral. Is your human need for approval, for eliciting likes and retweets, driving you to share questionable material? Does the content convey how you truly want to be perceived? You should have zero expectation of privacy when it comes to cyberspace.

Thanks to screen capture, even a deleted post can be preserved forever. This holds true even on platforms like Snapchat, whose core promise to users—that shared images will quickly vanish—has been known to fail. Just ask the nineteen-year-old day care worker from Mesa, Arizona, who posted to her Snapchat Story a photo of herself, seemingly giving a child the middle finger, along with the caption "SWEAR I LOVE KIDS!" While these were supposed to expire after twenty-four hours, is it any surprise that she was terminated after parents saw the photo and brought it to the attention of her bosses?[26] Understand that once something is posted, there is no rewind or delete key.

+ **Sharing with the wrong people:** You should frequently review the settings on your social media accounts and make sure you actually know who are connecting with. Who's in your Facebook friends and cell phone contact lists? Do you actually know them? Would you be embarrassed if you accidentally butt-dialed one of them? In 2010, Jimmy Kimmel dubbed November 17 National Unfriend Day, a time to review your contact list and weed out your true friends from your virtual acquaintances.[7] Just because you've set your privacy settings as high as possible doesn't mean you are 100 percent secure from trolls or a friend turned foe. You may believe that you're only sharing this with your core group, but remember, you don't always have control over what photos others choose to take and share.

Notoriously, back in 2012, a photo almost destroyed the life of Lindsey Stone, who was helping lead a tour group of developmentally delayed adults around the various landmarks in Washington, DC. When they stopped at Arlington National Cemetery, her coworker snapped a photo of Lindsey irreverently flipping off a sign that read SILENCE AND RESPECT. It was an ongoing goof between Lindsey and her friend, who thoughtlessly uploaded it to her Facebook page, tagging Lindsey. A month passed, and then, possibly because the privacy settings on those mobile uploads had been set to public, the image leaked out to enraged veterans, who created a Facebook campaign and got Lindsey fired.[8]

- **Sharing in haste:** People often refer to the phrase, "Think before you post." I say, *"Pause."* It only takes a second to post—and 60 seconds to pause. Take that minute to consider that post before you hit send. Picture yourself in that photo or receiving that email. Is this something that could be embarrassing or humiliating at a later date? Does it reveal too much information? Always ask the permission of others who are in the photo, especially with children, before posting it, and never assume that they have given you permission unless they have. If we've heard it once, we've heard it a million times—*think before you post*—but that hasn't stopped many of us from making digital blunders.

Sierra McCurdy sadly found that out the hard way with a simple emoji slipup. One evening, as her fingers were quickly tapping away on her smartphone, she commented on the day's tragedy—fallen police officers. Like most people, she was saddened to read the news and posted the following:

2 police officers was shot in Hattiesburg tonight!
- GOT EM'.

"They took it the wrong way," McCurdy said, explaining that the second part of her Facebook comment referred to the suspects, not the cops, and that she had meant to use the crying emoticon, to send her condolences.[9] This is the perfect example of why I recommend implementing *pause*. Pausing gives you actual time to stop and breathe. Furthermore, if you are considering sending a sensitive email, wait a full twenty-four hours and think it over.

+ **Sharing without dignity:** When we see adults, politicians, celebrities, or athletes acting childish or bullish online, it sends the wrong message to our fellow adults and to our kids. Many of these people are role models who our youths look up to. But when we have videos circling of hip hop-stars sniffing cocaine over a woman's breasts and politicians trashing the reputation of private citizens or getting caught with their digital pants down, like Anthony Weiner, over and over again (and over yet again), we have crossed a line.[10]

There's a difference between being *clever* and being *cruel*, and sometimes those lines are blurred, especially online. Celebrities may think it's a fancy way of telling a joke, or a sports figure may believe he's sharing cool lingo, but keep in mind, the translation on the Internet is not always the same as it would be off-line. We must learn to respect our language online, whoever we are. No matter your status, age, or ethnicity, if you have a keyboard, you are responsible for your *own* keystrokes. And keep it clean. There's never a need to use offensive language.

+ **Sharing with negativity:** I'm sure everyone knows people who uses their social media feed as a venting machine. The complaining never stops, whether it's their bleak life, their horrible job, or their dismal dating scene. Worse is when they impose their negative thoughts on your good fortune—you've just landed your dream job, and they make an unenthusiastic remark like, "Not a great company to work for." Yes, we've all experienced the Negative Nellies and Debbie Downers in our world, and we don't want to be one of them—especially online. From the moment you are given the privilege of your first keyboard, your virtual résumé begins. It's up to you to maintain and create a positive persona. It's true, we can't be happy all the time, and it's fine to reach out for support in times of grief. But the good news about the Internet, and even your smartphone, is that you can turn it off if you're having a bad day. Also, never post something in haste or anger that you might later regret—log off instead. I like to say, "When in doubt, click out."

OFF-LINE MATTERS TOO

Sadly, how we behave when we are out and about in the world can now be captured and posted online to destroy our reputation. How often have you seen someone losing their temper at a sales clerk, at the DMV, or at a flight attendant? Celebrities have long known the risk of misbehaving in public, but now, even noncelebrities can be subject at any time to someone flicking on their smartphone camera and filming. In 2008, Barack Obama's newly appointed head speechwriter, Jon Favreau, then twenty-seven years old, found himself in a tricky situation after goofing around at a party with a life-sized cardboard cutout of former rival Hillary Clinton, the incoming Secretary of State. He mockingly posed for a photo, placing a hand over her breast while another staffer raised a beer bottle to her lips. Weeks later, images of him "groping" her were uploaded to Facebook, prompting news coverage in the *Washington Post* and beyond, forcing Favreau, now a political podcaster, to offer a behind-the-scenes apology.[11]

Being Cybersavvy

The good news is there are concrete ways to reduce the risk of many of these scenarios happening to you. You have your annual exam to take care of your physical health, but when it comes to your digital wellness, your online checkups need to be more frequent.

Regularly reviewing your virtual exposure should be a priority. From getting into the habit of assessing your privacy settings on all your social platforms to decluttering your address books to being aware of social media policies at work or school—your virtual checkups can prevent headaches later.

Let's review:

+ To start with, as difficult as this can be for many people who have been victims of cybershaming, perform a Google search on yourself regularly. This can help you become aware of how your name is being used online.

+ Create a simple Google alert that will notify you when your name is mentioned online. These alerts are typically great at letting you know when your name appears in a news article or sometimes in a blog post, but it won't necessarily catch every time your name is mentioned, especially in things like posted comments. So you still need to be diligent about keeping track of your name through your own regular Internet searches. When setting up the alert, be sure to put your name in quotation marks, and don't forget to create additional alerts for any variations of your name that you use too.

+ Make a habit of keeping up with your social media privacy settings, checking them and securing them often. Privacy settings have a tendency to change without notice, so losing track of your settings can lead to unexpected—and unpleasant—surprises.

+ Weed out those so-called friends on your social media accounts, the ones who you just don't know.

Distinguishing real friends from cyberfriends can really cut back on the chance of your personal information getting into the wrong hands.

+ Create targeted lists of friends, to make sure you're only sharing things with your innermost circle. Social media platforms such as Facebook and Twitter provide features for you to build select groups of friends and family, so you can share comments and photos with your intended recipients only.

+ Find out whether your college or employer has a social media policy in place and determine what's considered out of bounds according to its guidelines. If in doubt, ask your company's human resources department.

Being cybersavvy is the first step in maintaining your online real estate.

Claiming Your Social Media Profile

The best defense, as they say, is a good *offense*. The best thing you can do before a digital disaster strikes is stake a claim over your name and create a positive Internet persona. "Most people think that an online attack would never happen to them—or they underestimate just how severe it can be," says ReputationDefender's CEO, Rich Matta. "But it's like that old saying: an ounce of prevention is worth a pound of cure."[12]

Own your names on Twitter, Facebook, and LinkedIn, even if you don't want to build up that presence right now. Why would you let someone else with a similar name take the website, Twitter handle, or Facebook page that makes the most sense for you to own? "It's always easier to combat the duration and severity of an online attack if you already control the sites at the top of your search results, your social accounts, and other online assets prior to the attack," Matta explains. "These sites alone give you the potential to control three of the ten positions on page one of your search results."

Next, begin creating positive information about yourself. Teens and young adults should use their social media accounts as an asset, creating LinkedIn profiles or Twitter feeds that will impress college admissions officers or future employers, says Alan Katzman, founder of Social Assurity, which has coached nearly a thousand high school and college students on this technique. "You have to learn to post content that won't generate likes or follows from your group of friends, but toward your future audience, who will [use it to] try to determine who you are," he says, in reference to his clients' potential employers and college recruiters.

For many young people, the problem is not necessarily wiping clean a social media profile littered with red Solo cups and bikini selfies, it's simply a lack of anything impressive—like community service or academic accomplishments. "It's void," Katzman says. "You're not telling your story. That's the biggest problem—they're looking at nothingness." One of his clients created an Instagram account depicting her caring for horses and riding competitively, making sure that the college admissions team received the link along with her essay. "It's telling your story in the digital world and putting it out there," Katzman says. "That's all you need, really, that 'social proof,' that 'I'm not writing this essay to impress you, not conjuring up this image. This is me.'"

Claim your domain name, which you can buy from sites like GoDaddy or register.com, and build your own personal website, which can be simple with free services like Wix. Here are a few tips from the online reputation management firm BrandYourself on how to strengthen the power of your site.

* A .com extension is preferred to alternates such as .tv; since .com is more widely used, it ranks higher in search results.
* Registering for as long a period of time as possible (five years versus one) adds to your credibility in search engines' eyes.
* Create a Bio or About Me section, using your full name as a keyword.
* Activate as many social media accounts as possible, like Twitter, Facebook, LinkedIn, Google+, Pinterest, Instagram, Tumblr, Avvo (for lawyers), etc., fill out your profile for each account, and link them back to your own site.
* Feed your social media profiles through your site, and create blog posts to keep your site content fresh.[13]

Once you've surrounded your online persona with positive information, one small negative comment, say, a blog post by a spiteful ex-employee, won't be able to drown out the good.

If you're a small business owner, ReputationDefender suggests:

- Respond cordially and promptly to online reviews.
- Generate more reviews from a broader and more representative set of your customers by asking all of them to review you online.
- Take disputes off-line as quickly as possible. Be pragmatic with refunds and other ways to "make it right" for the customer, even if they're wrong.[14]

Scrubbing Your Past

Like our Miss Teen USA discovered, social media postings from long ago can get you in hot water later if you land in the public eye. But you don't have to be a beauty queen for this to happen. What if you're just looking for a summer job? One mother interviewing a potential nanny to watch her children discovered an inappropriate video buried four pages deep in a Google search. Do you think she hired that young lady? Of course not. But she never said a word to her about *why* she chose not to hire her—deliberately leaving the video lingering in cyberspace for other potential employers to discover in the future. Checking out someone's earliest social media history even has a name: backstalking, or scrolling back to someone's very first Facebook post or album, to when they had that bad middle school haircut and braces (and, if so inclined, even posting a comment so that the photo leaps back into their present-day newsfeed, front and center, potentially embarrassing them).

Social Assurity's Alan Katzman believes that 99 percent of

the embarrassing things you post at age fourteen or fifteen will have no significant impact on your academic future. "But," he adds, "there are a few radioactive posts—if you come off as being racist, if you talk about hate or violence. Colleges want to assemble a safe, diverse, open community. It's okay to have an opposing point of view, but it can't be couched in intolerance." One strategy some sneaky students are using is to change their names on Facebook during the college admissions process, so that their history is harder to find, even though the platform has policies against doing so.

Try thinking back if there are any abandoned or embarrassing accounts of yours lingering in cyberspace. Do you still have angst-ridden poetry up on Myspace? Rap lyrics you've quoted that might be misinterpreted by more mature audiences? A Facebook album of one hundred filtered shots of you taken with Photo Booth? You may want to backstalk yourself and consider setting those earliest images to private. You can also visit JustDelete.me, a directory of direct links to your abandoned accounts from various web services, from AOL to Zynga, so you can delete them.

BrandYourself offers a basic scan of your Google search results, plus all your social media history on Facebook and Twitter, that will turn up controversial or damaging posts. It's not a bad idea to run an image search on your name as well. Toni Birdsong, a family safety evangelist for McAfee, recommends looking not only at what you've posted, but even things that other people have posted publicly that you've liked. "Others could judge you guilty by association," she writes. "It may be time-consuming, but you can clean up your Facebook 'like' history in the Activity Log."[15]

Protecting Your Present

It's always wise to shore up your privacy. Is your personal information scattered all over the web? Real estate records are public, so it's hard to protect your home address from being listed without resorting to extreme measures, such as setting up a trust and a post office box. However, you can pay to keep your home phone number unlisted. Lindsay Blackwell, an antiharrassment researcher, recommends using a Google Voice phone number that forwards to your personal phone number but can later be disconnected, for added security.[16] You can also keep the contact information on your website private by purchasing domain privacy when you create a site. Once this information has been released, it is difficult to get it back, so consider it a small investment in the future.

Always be aware of where your contact information lives on the Internet. Have you uploaded a copy of your résumé onto your website that includes your home address? In some states, boating records or voter registration lists are published online, unless you request otherwise. You can also opt out of online white pages like Spokeo, Pipl, and PeopleSmart by filling out a request, although it can be a Sisyphean task—there are more than two hundred such data-collecting services.

Do you go off-line when you're away from home? Security expert Theresa Payton advises against using free Wi-Fi, because you don't know the security team or the protocols in place. "I equate free Wi-Fi to using a toothbrush you found on the ground—you wouldn't use that because it's free, right?" she asks. Instead she recommends using your smartphone or a portable Internet hot spot as your Internet connection. "If you find you must use free Wi-Fi," she adds, "ask questions about the security of the signal and be wary of clicking on links or pop-up ads."[17]

The antiharassment website Crash Override Network offers an automated cybersecurity helper, which can guide you to safeguarding your system. Some of its best tips to avoid hacking: use an app like Prey or Lookout, which can allow you to remotely wipe your phone if it were ever stolen, and never download unknown attachments. They also advise improving your password protection with password managers like LastPass or KeePass, and shoring up your own website from automated attacks. Have you wondered why password recovery questions have gotten strangely specific? (One site, for instance, wanted to know the color of the towels in your master bathroom.) One reason is that asking questions about your high school mascot or the street you grew up on are things that a potential hacker could easily deduce with some cursory research.

Another antiharassment group, CommunityRED, advises that you can also deter hackers by changing the security code on your iPhone from four numbers (don't leave it as 1234, please) to something much more complex. It's as easy as going to your settings and turning off "simple passcode." Also, have you set up your two-step verification to protect your email account? This requires entering a second form of identification, such as a code you receive by text to your cell phone, whenever you are logging in. Finally, cybersecurity experts recommend visiting the site haveibeenpwned.com, to see if any of your accounts on sites like LinkedIn and Myspace have been associated with a data security breach.

A Simple Band-Aid Solution

Remember teen beauty queen Cassidy Wolf, who was compromised by a hacker who hijacked her own computer's webcam? It's not that hard for those with criminal intent to infect your computer with malware that turns on the camera while disabling the light that is supposed to alert you when the camera is running. The Brookings Institution report on sextortion, discussed in chapter 2, recommended that webcam manufacturers consider ways to improve the security of webcams—by providing slipover masks or even a physical on/off switch so a user can ensure it is disabled.

Until they do, the basic ways to try to avoid malware are as I've previously stated: never download unknown attachments and avoid using unsecured Wi-Fi when out in public. But there's another low-tech, easy solution to this particular problem: take a Post-it note, a sticker, or a piece of tape, and cover up your webcam whenever it's not being used. You can even use a Band-Aid. I'm not kidding.

The *Guardian* reported on a photo taken of Facebook creator Mark Zuckerberg that seemingly shows that even he covers up his webcam with a piece of tape.[18] And former FBI director James Comey reportedly told a crowd that he does the same. "I put a piece of tape over the camera because I saw somebody smarter than I am had a piece of tape over their camera."[19]

You should do the same.

Public and Permanent

The Internet is *public* domain. Did you know that the Library of Congress is documenting every single public tweet that has ever been made? Sites like Snapbird.org allow you to search old tweets going back much further than the Twitter search engine currently allows. Even old versions of websites that you redesigned ages ago are still viewable, thanks to the Wayback Machine (web.archive.org), an Internet archive that crawls the web and preserves blasts from the past. Take a moment and search your own website (if you have one) to see what information lingers online.

Know that everything you put out there has the possibility of becoming "Public and Permanent," an expression perfectly coined by Richard Guerry, founder of the Institute for Responsible Online and Cell Phone Communication. "Far too many people with technology are not stopping to think about the long-term repercussions of their actions," he says. Guerry advocates for digital consciousness—always posting with the awareness that anything you've documented could be disseminated. "There is no way to control what is going to happen, none," he says. "Digital tools were never designed for privacy. We're going against the grain for what these tools were intended. By no means is everything going to be Public and Permanent, but you have to be prepared. Think about your legacy. It's not just imagining [that] your ninety-year-old grandma will see your naughty text—but [that] your own grandkids will too."[20]

Even with the strictest privacy settings, we don't have control over human behavior, technology glitches, or cybercriminals. Maybe your husband leaves his phone behind on the train or it gets picked up in the locker room. Maybe your friend forwards your nasty text to a mutual friend. Maybe your Carbonite or Yahoo account has been compromised—how many times have you received an apologetic email from a tech company saying that your online password and sign-in information may have (oops) been accidentally divulged?[21] And don't expect others to respect your family members' privacy. A recent Nominet study of two thousand parents found that, on average, parents have uploaded a photo of *someone else's child* nearly thirty times in the last year, and 36 percent don't bother to ask permission to do so.[22]

Whenever you digitally document anything—anywhere—you need to realize that there is a distinct possibility of it becoming forever engraved online. Think of it as writing with a Sharpie, not a pencil. Assuming that there is no risk of a third party finding it is nothing short of ignorance. Privacy settings can help, but they're not always reliable, so we must start with common sense. You never know what one person will share with another, then with another. "When posting on Facebook, even the strictest privacy settings do not prevent posts you share or like in a public setting—such as public groups, friends' cover photos, business pages, and some events—[from] being viewed by others," explains Australian cybercrime prevention specialist Janita Docherty, founder of CyberActive Services. "Comments and photos you post to these areas can be viewed by your friends, your friends' friends, and the wider Facebook community. These posts can go viral very quickly. A quick screenshot or share from a public post and it's on its way."[23]

Pause. Reconsider what you are about to post or send into cyberspace.

- Is it an embarrassing photo?
- Are you using profanity?
- Are you posting a photo of someone without his or her permission?
- Are you disclosing personal information (address, phone number, etc.)?
- How would your friend/employer/potential client react to this post?
- How would you feel if this post were made about you?

"I look at technology as a twenty-first-century flame," says Guerry. "It opened opportunity, [but] use it the wrong way, [and] it's going to hurt." He likens smartphones, Snapchat, and Twitter to putting a matchbook into the hands of children without teaching them fire safety until *after* they've burned down the school. "The first digital generation is always going to pay the price," he says. "We are creating statistics for our children's children to learn from."[24]

Noods

As we just learned, everything is permanent online. If someone asks you to send them a nude photo, my advice is:

Run.

Just because it's frequently done doesn't mean it won't land you in serious digital consequences. Slut pages aren't only a child's playground. In 2017, it was exposed that some U.S. Marines were involved in creating social media pages full of nonconsensual shared nudes of their colleagues.[25]

Sexting isn't just common among swinging singles and digital natives. One recent study in *Cyberpsychology, Behavior, and Social Networking* found that 12 percent of married couples admitted to sending nude or nearly nude photos to each other.[26] However, with the divorce rate of first marriages at forty percent and of second marriages at sixty percent, the impulse to get even with your spouse has taken a new, evil turn. Another study found that 4 percent of online Americans, or 10 million men and women, reported either being threatened with revenge porn or actually victimized.[27] If you're considering taking or sending a nood, make sure you are aware of the risks and ready to deal with the potential consequences. What could the impact be on your future relationships? Employment? Career?

Personally, I don't believe the risk is ever worth the reward.

Even if you trust your partner, do you have faith that he or she will never share your images even with their best friend? One survey revealed that one out of three adults will share each others' sexual images without their permission.[28] What if the cloud-based storage you use is hacked? If the worst does happen, it's important to remember that it's not your fault—you trusted someone with private content, as you have a right to, and it's the criminals who share that content publicly without your consent.

If the photos do get out, either from being hacked or posted intentionally, "It can really be devastating professionally," says attorney Christina Gagnier, a board member of Without my Consent, an advocacy group that helps victims of nonconsensual pornography and other online harassment. "I have spent thousands of hours working for victims taking content down, because once an image is up, it just spreads like wildfire. You have to really twist arms, and get restraining orders and injunctive relief from a court. There's a lot of steps you have to go through, and the average person doesn't have the financial resources or the time to be getting this content down and using the legal system to do so."[29] Sadly, some unscrupulous sites will post photos and then demand thousands of dollars from the victims for the supposed "costs" of removing the content.

You also have to ask yourself, will sending that nude image really help your relationship? Not necessarily. According to Michelle Drouin's research published in the *Journal of Sex Research*, women who are insecure about their partner are the

ones most likely to sext images—and doing so doesn't help solid-
ify the bond.[30] "There is no research that shows this is beneficial
to a relationship," says Drouin. "I don't think it will accomplish
what you want." Her advice instead? "Have a meaningful conversation, actually *have* sex instead of sending a sex message."[31]

For teen girls, the odds of a sext landing a serious boyfriend
are even more far-fetched. In a 2015 survey, sexting researcher
Elizabeth Englander, PhD, of the Massachusetts Aggression
Reduction Center, found that of the students who were pressured
into sending a sext hoping to land a potential boyfriend, only 2
percent managed to do so.[32]

How can we help our teens, like those in Duxbury, resist this
pressure? And how can we talk about why this is not an acceptable
request? If you are underage, the possession of nude photos, even
if the photo is a selfie you took of your own body and on your own
phone, can often be considered child pornography. While most
prosecutors have restrained from charging teens swapping selfies
as sexual offenders, it can happen. Two teens, a sixteen-year-old
North Carolina high school quarterback and his girlfriend, found
themselves in exactly that situation, charged with the felony of
child porn after consensually sexting one another pictures.[33]

Since many states are reforming laws to reduce this to a
misdemeanor, teens may not know the law but they generally
know that they and their friends are all sexting and none of them
are going to jail or being registered sex offenders for it, Englander
says. But the negative consequences are real. She found that of
the students who felt pressured to send a sext, only 39 percent

later reported that it caused no problem, while a quarter said it made them feel "worse."[34]

Some will argue, ignoring the inherent risks, that teenage sexting is relatively harmless, that teens need a safe outlet to explore their sexuality, free from the risks of STDs or pregnancy. Professor Drouin believes that we adults need to provide teens with an out. "What teens need is help with what to say when someone asks for a picture, without jeopardizing their relationship," she says. "That could be saying, 'I'm not ready yet,' or 'I'd rather keep our physical [experiences] in person,' or simply, 'My parents will confiscate my phone if they find out.'"

Emily Lindin, author of *UnSlut: A Diary and a Memoir*, has a strong message that always strikes a chord when she talks with teen girls about slut shaming: she points out that online porn is readily available, so these boys already have all the masturbatory material they could ever desire—what they are really after is power to lord over you, control you, even blackmail you. She asks young women pointedly, "Do you really want to give them that power?"

Emily's question can equally relate to adults. Ultimately, the decision to share nude photos of yourself is yours, and yours alone, to make. And if it happens that someone out there chooses to exploit your nude photos—*it's not your fault*. But, please, before sharing a nude, know the risks and take time to consider the potential consequences.

Sharing Selectively

When we were young, we were taught that sharing is caring. But sharing has taken on a new meaning when it comes to the digital world. From parents posting their kids' images online, despite the risks, to braggarts posting their elaborate vacation photos and job promotions—social media has become the place to document your life. A place where people pretend to be someone they aren't, a place where they can have a lot of virtual friends, even if they don't have many actual ones.

But what happens when you have that one bad egg? The one who is a bit off-balance, maybe envious of your life?

Hurt people will hurt other people.

Isn't that the truth? Misery loves company. That's when oversharing can sometimes get you into trouble.

It's about knowing your *social liability*.

You may be tempted to post a reaction to an event, respond to a photo that makes you chuckle, or even comment on something you don't agree with. No one is saying you have to stop living life or expressing your opinions online, but you need to use your digital wisdom before sharing your keystrokes, photos, emojis, text messages, and emails. With the Internet, it's very easy for your message to get lost in translation. A simple line can be misconstrued, and soon you could be facing cyber-bullets and an online attack for something taken completely out of context.

Learn to be selective in your sharing process. Be aware of who your friends and cyberfriends are, especially those who are your coworkers. The fact is, if you find yourself questioning a post, it probably doesn't belong online.

For cybersecurity measures, follow these guidelines for what kind of information should and should **not** be shared on social media platforms.

Never Share

+ Personal financial information
+ Address and phone number (especially landlines, which can be reverse-searched to find the associated address)
+ Social security number
+ Birth date
+ Your kids' pictures on public pages
+ Private, sensitive information or gossip

Share Limitedly

+ Kids
+ Family
+ Pets
+ Details about your place of employment
+ Vacation plans and photos, at least until you have returned from your trip

+ Celebrations in your life, such as marriage, birth of a child or grandchild (use your privacy settings for selective friends and family)

Share More Openly

+ Articles of interest
+ Great quotes you like
+ Books, movies, and other media you like
+ Hobbies, interests, and volunteerism
+ Accomplishments (educational achievements, professional promotions, awards)

Tending to Your Electronic Legacy

Email Etiquette

Snafus involving email have been widely written about, but I want to touch on some of the most important points that I believe can frequently lead to cybershaming if not avoided.

First, stick with your given name for your email account. Gone are the days when *imhot4u@dot.com* or *suzeluvsmickey@dot.com* were cute. With email addresses like these, you risk your résumé or college application being passed over. Human resources managers and business owners say that when they see such cutesy email creations, they swiftly discard the application. Is this fair? Perhaps not, but it's life today in the digital space.

Parents should also create email addresses for their children in their full names, for use when they're older. It's also time to stop with the adolescent slang for handles and usernames. No more *HotLipz ILuvJesus SexEMama ChillNBeanz KatchngWaves*.

The tone of your emails matters. USING ALL CAPS doesn't demonstrate excitement; it's considered screaming. Use moderation with your emotions and don't overuse exclamation points!!! There are times when time-honored traditions are warranted. Keep your grammar and spell-check *in check*.

Be cautious with using humor in emails, as this can easily be misinterpreted. Without seeing someone's facial expression or hearing their tone, it's hard to determine whether they are joking or serious. The fact is, emails have become a place where longtime friendships can break down and end due to miscommunication. It's important to remember that email is, in many ways, still only mail. If you want to have a relationship with someone, you need to continue to nurture it face-to-face.

Sending off an email today is risky; a simple wrong click and you could face your private message being spread globally or misconstrued. Shaming has now become only an email away if you're not careful with your digital skills.

Sorry, Wrong Number?

My grandmother used to love writing me long letters. I still have many of them today. The only etiquette she worried about was that I wouldn't be able to read her writing or that her content would bore me. Neither was true.

Today, with email and texting, the concerns can be much more serious. Even at ninety-nine years old, my grandmother was never at risk of negative repercussions from sending that letter to the wrong address. With modern forms of communication, however, it can be a completely different story. How many duplicate first names do you have in your contacts list? Probably several.

Have you ever sent a sensitive email or text to the wrong person? A person you barely know, who is only in your contact list by acquaintance? You are typing fast, as we all do, the name autofills, and it seems like it's the same Mary you email every day—but suddenly, right after you click that mouse to send the message, you realize it's not.

Hopefully, you've sent the email from a Gmail account and had already set up the "undo send" functionality, which at least gives you a very short period of time—up to thirty seconds—to recall a mistaken message. But what if you haven't? If you've ever seen Ellen DeGeneres's "Clumsy Thumbsy" segment, it can be hysterical. But not so much when it's happening to you.

One attorney meant to send a confidential text about his celebrity client, a professional athlete privately struggling with drug issues, to a fellow attorney. Instead, he sent this message directly to the Associated Press:

Heaven help us if one of the conditions is to pee in a bottle.

Soon after, that embarrassed lawyer was no longer representing this particular sports star.[35]

One way to prevent situations like this the next time around is to take a closer look at your contact list.

Who exactly is on that list? Do you know them—personally? You may have someone in your cell phone contact list who called you once or twice, but who you wouldn't want receiving a pocket call from you or a sensitive text message.

But what if you're like me, and you don't want to delete that person, on the off chance that you need to get in touch in the future? The answer: open the contact card of the person in question, copy the email address and phone number into the *memo section* of the contact card, then delete the email and phone number fields. This will prevent your device from automatically pulling up the email address or dialing the phone number if you accidentally select that contact, and this person who rarely hears from you won't be suddenly surprised by your email or number showing up on her screen. This strategy is also ideal for preventing spider spam emails—emails from someone in your contact list who is unaware that she has been hacked.

Finally, when you're typing in a phone number for a new contact and sending a text with sensitive material, beware. One twenty-nine-year-old Florida man looking to score some drugs tried to text a friend for a hookup. Instead, he accidentally texted a narcotics captain in the Martin County sheriff's office, likely by typing in an incorrect digit in the phone number. He was eventually arrested for drug possession, and the sheriff's department posted the entire text exchange on its Facebook page, under the heading WHEN TEXTING GOES WRONG.[36]

The Reply-All Oops!

Another major hiccup that is preventable is the reply-all oops. Have you ever received a group email or text and intended to respond to only the sender about someone in the group? Maybe you thought that person shouldn't have been included? Perhaps you even said some distasteful words? Then you realize that you replied all, allowing everyone to witness your snarky remarks and making a new enemy!

Whether you're group texting or emailing, the same guidance applies. Always double-check your recipient—be sure that you are sending your reply only to the one person who is supposed to receive your message, and no one else. With texting, you need to start a new message completely. Unlike emails, you rarely know whose phone numbers are in the group, so it's important to have your digital wisdom on an all-time-high alert.

Finally, please remember with email, as with anything text-based, there is *zero* expectation of privacy online. We all witnessed the damaging email hacks of various Democratic staffers during the 2016 U.S. presidential election, shared by WikiLeaks. Once your email is sent, even if you have confidentiality clauses at the bottom, you never know who will intercept your information.

The Art of Commenting

There's a good reason for the saying "Don't read the comments." Too many of us have found that online comment sections have deteriorated into sludge-throwing slugfests between anonymous cranks. Even online obituary sites, like Legacy.com, have been forced to hire staffers to screen comments and flag cruel words—against the deceased.[37]

It's no wonder so many major news media sites, like Huffington Post, have put an end to anonymous postings, and NPR, Reuters, and the *Chicago Sun-Times*, among others, have done away with comments altogether. Why? Part of the reason was because adults with a keyboard couldn't behave themselves online. "We got a lot of trouble in our comments on different stories—attacks on our writers, just stupid things; it wasn't smart," Recode's executive editor Kara Swisher told the Nieman Lab.[38]

However, the fact is, there is an art to commenting smartly, and smart comments can help your own online reputation. Remember, everyone is reading *your* responses, so it's a great opportunity to make an impression on others. Depending on your remarks, they may look forward to reading more about you or learning from you.

Just because we have the privilege to leave a comment doesn't give us the right to abuse it. Sometimes we forget that there's a live person behind the actual image or story who is reading the comments (not to mention counting the likes). What you say counts and needs to be done with care. It goes back to the old cliché that has been taught for generations: "If you don't have anything nice to say, don't say it at all." Given today's cyberlife, we need to take this advice with us into the digital world.

There isn't anything wrong with making comments express-ing a different opinion about a topic, but intentionally causing harm or knowingly inflicting pain unto someone should never be a reason to leave a remark. Or, as I like to say, be constructive, not combative.

Despite all your efforts to be a positive, conscientious commenter online, not everyone will treat you with that same respect. In the next chapter, we will look at how some people handle the online hate and nastiness of others and what we can learn from it.

CHAPTER 7

TO FEED OR NOT TO FEED

SIX WAYS TO RESPOND TO TROLLS

"We cannot continue to allow Internet mobs to stifle the free speech of others."

—Beverlee J. McClure[1]

Diffusing Digital Drama

Don't feed the trolls.

That's long been the best—and only—advice for those dealing with online harassment. Is that really the ideal way to react?

First, let's clarify the type of attackers we're talking about. Are these recreational trolls, who are simply looking for sport? Or criminal trolls, who have a more malicious or harmful intent, which may force you to take a more serious approach? (We'll address the latter scenario in more detail in chapter 8.) Have they broken the law, posted a photo that invades your privacy, or made threatening remarks? Or are they simply mocking or criticizing you, as hurtful as that may be? Lindsay Blackwell, online harassment researcher at the University of Michigan, points out that in its original online context, the term "trolling" described the use of humor to poke fun at or rile up clueless newbies on a particular platform. "Classic trolling is funny and harmless," she says. "Its goal is to draw attention. It is unpleasant but doesn't escalate."[2]

California Internet attorney Mitch Jackson explains the distinction this way: "Recreational trolls [are] annoying people who, on occasion, end up in your social media feed and say stupid things. Ignoring or blocking [them] will usually remove recreational trolls from your social media life. Criminal trolls [are] the type of trolls who are intending to harm you, your family, or your business. Ignoring or blocking [them] will usually only result in the trolling getting worse. It's the difference between walking down the street and stepping on a random piece of chewing gum (recreational troll) [and] having someone take an old and well-used piece of chewing gum out of their mouth, stick it on the end of a baseball bat, and hit you with it as you walk by every single day."[3]

Today, of course, much of online trolling has taken a far darker turn. Kate Bigam, a twentysomething blogger from Ohio, retweeted a video from a man who claimed he was harassed for speaking Arabic on a flight. The story turned out to be a prank, but her message drew attention from white supremacist trolls, who deduced that she was Jewish and began bombarding her Twitter account with anti-Semitic messages. "I was, frankly, just blown away," she says. "You never expect, when you're banging out a quick 140-character commentary on the world, that it's going to result in a massive assault. When the tweets started coming in, I started crying right there, trying to hide it from the people sitting around me. It was just so overwhelming and hateful."[4]

Assuming that your trolls are recreational, one strategy is to let others speak up on your behalf. *Scary Mommy* blogger Melissa Fenton generally takes a hands-off approach when nasty, tasteless, or horrific comments pour into one of her blog posts or her Facebook feed. "Arguments will start in the comments threads, [and] I let them work it out," she says. "I have regretted the time[s] I have gone back arguing with someone. You're not going to change anybody's mind, and it's not worth it. People who have that much hate and [who take] joy in stirring up discourse and want to cast stones, there's some kind of deeper issue going on." But she has noticed that online communities are going to bat for each other. "People are really starting to stand up for people. Because you never know when it's going to be you."[5]

In some cases, if what triggered the attack is something you've done, removing the original offending post or tweet is a good first step to stop the bleeding. "I did delete the original tweet when I realized it was a sham," Bigam says, "because I didn't want to perpetuate fake news." She also made her accounts private and began blocking and reporting the abuse. "The whole thing went on for about forty-eight hours, and even weeks later, I sometimes get random tweets from an alt-right account who's late to the party but still wants to share with me some anti-Semitic meme."

In many situations, no matter how many times you explain that the sky is blue, there will be people who believe it's yellow. If you feel you must wade into today's cesspool of contentious political, racial, or religious online debates, all I can say is brace yourself for possible pushback. It can be like attempting to put out a fire with a jug of gasoline. One young adult novelist found her upcoming book, which hadn't been released yet, trashed with one-star reviews on Goodreads after she engaged with anti-Semitic and white supremacist trolls on Twitter.[6] Social activist Suey Park, credited with creating Twitter hashtags like #NotYourAsianSidekick to highlight Asian tokenism in Hollywood, suffered a misfire when she took on a joke posted by *The Colbert Report* on its official Twitter account. Her critique, calling on her followers to "#CancelColbert—trend it," sparked a backlash of vicious taunts and death threats from users on 4chan and Reddit, forcing her to cancel speaking engagements and flee her home. Twitter users told her, "Go get raped," and, "You should probably just kill yourself."[7] In certain cases, engaging with trolls can even put your friends, family, and innocent bystanders at risk of being targeted themselves. Incredibly, when Harvard under-grad Lena Chen's blog, *Sex and the Ivy*, came under attack, even people who had simply been reading and posting comments on her site began to experience online harassment of their own (see chapter 8 for more on this).

Strikingly, without fail, everyone I interviewed for this book, from Emily Lindin to Lindsay Blackwell and others, told a personal story about how they found themselves on the defense from trolls just for stating their beliefs. "I've experienced people taking shots at me, question[ing] the way I look, the way I talk," says Silicon Valley tech journalist Larry Magid. "Anytime you do something that might be provocative, or courageous, you do increase the risk [that] people will lash out at you. One way you could potentially minimize it is [to do] something you shouldn't necessarily have to do—keep your head down. Everyone should say what's on their mind, but you should be aware, you are taking a risk. There are times when I choose to hold my tongue, to not say something that I might otherwise say, because I just don't feel like dealing with trolls."[8]

If you are determined to take on your cyberthrashers yourself, and feel mentally prepared to do so, Mitch Jackson offers some methods to address these pesky trolls (but remember, some of them will see this as entertainment):[9]

+   Call attention to them without specifically engaging them.
+   Ask the troll to fully identity himself or herself, and share his or her full name, email, or website.
+   Avoid emotional arguments, and only use facts.

In general, of all the strategies to consider when dealing with purely recreational trolls, my personal advice is: don't engage. It will only fuel the drama. But enduring it silently might not feel right. You may decide that you want to stay and fight back against your critics, or to see if you can reason with them. Might you be able to charm the blogosphere over to your side? Let's look at some examples of how several vastly different approaches to handling cyberdrama played out, and see if you can find a strategy that best fits your situation and your style.

The Activist

> *"Overnight, I went from being a completely private figure to a publicly humiliated one. I was Patient Zero."*
>
> —Monica Lewinsky[10]

Even before the existence of iPhones, Google, and Facebook, Monica Lewinsky became one of the first victims of global online shaming. Like many who suffer cyberhumiliation, Monica fell silent for a long time, but today, she's an inspiration fighting back against the worldwide chorus of nastiness.

Flash back to 1995: a young Monica, fresh out of college, was excited to start a career in politics as an intern at the White House. She could never have imagined the roller coaster she was about to endure. Workplace romance is not new—but when you are employed at the White House and your boss is the president of the United States, that relationship is taken to a whole new level. Let's be clear, it takes two to tango, but Monica's youth had yet to catch up with the wisdom that comes with age and experience. When the news of her relationship with the president was revealed, she was devastated to find herself branded a tramp, slut, whore, tart, bimbo, floozy, and even spy. Monica was shattered.

"What does it really feel like to watch yourself—or your

name and likeness—to be ripped apart online?... It feels like a punch in the gut," she recalled. "I came close to disintegrating."[11]

After more than a decade of silence, Monica has now become a force to be reckoned with, championing this cause that she personally understands is strangling our culture. In 2014, she began to speak publicly about her mission to end online abuse at Forbes' 30 Under 30 summit. "The experience of shame and humiliation online is different than offline," she noted in her speech. "There is no way to wrap your mind around where the humiliation ends—there are no borders. It honestly feels like the whole world is laughing at you. I know. I lived it."

Shortly thereafter, in 2015, she was invited to present a TED Talk, "The Price of Shame," which has received more than ten million views. As Monica says in her opening, "When this happened to me seventeen years ago, there was no name for it. Now we call it cyberbullying and online harassment." She continues, "Cruelty to others is nothing new, but online, technologically enhanced shaming is amplified, uncontained, and permanently accessible... The echo of embarrassment used to extend only as far as your family, village, school, or community, but now it's the online community too. Millions of people, often anonymously, can stab you with their words, and that's a lot of pain, and there's no perimeters around how many people can publicly observe you and put you in public stockade."

Monica is forging forward, making up for lost time and using her voice as an international speaker and social activist. She is an ambassador for BystanderRevolution.com, and in collaboration with Vodafone Telecommunications, helped design emojis that are specifically for people to use when they witness cyberbullying or for someone that needs support. These unique heart-shaped emojis, some with hands reaching out in a cyberhug, are perfect to send when you can't find the right words.[12] As a regular contributor to *Vanity Fair*, her articles are written with true compassion, eloquence, and insight. Defying her critics, Monica continues to speak out, hoping to protect others from what she calls being "humiliated to death."

"I want to put my suffering to good use," she says, "and give purpose to my past."[13]

The Fighter

"Each one of those people is a real human being, a real person whose world imploded the day they found themselves to be a punch line on a giant stage."

—Caitlin Seida[14]

It was Halloween 2013, and Caitlin Seida had dressed up as her favorite video game character—Lara Croft from the video game and movie *Tomb Raider*. She posted it on her Facebook page, neglecting to check her privacy settings so that the photo would be visible to friends only.

It was not long after when a helpful friend sent over this alarming note: "You're Internet famous!"

Caitlin's photo had been copied and pasted and was now

hurtling through the Internet on Tumblr, Reddit, 9GAG, FAIL Blog, and beyond, with the words FRIDGE RAIDER superimposed across it. At first, Caitlin says, even she had to chuckle at the clever wordplay. But then she started reading the hateful comments posted below—many from other women. "Heifers like her should be put down," was one among hundreds of remarks, referencing her heavy frame, which is due to polycystic ovarian syndrome and a thyroid condition. Sure, there were a lot of defenders. But the hate was inescapable. "The world imploded—and took my heart with it," she wrote movingly in a piece for *Salon*.[15]

As she scrolled through, she realized that most of the comments were traceable to the women's own Facebook accounts. Caitlin decided not to just take the abuse, but to shame them back. And not in a tactful way. "You're being an asshole," she messaged them. "Why don't you just do the right thing and delete the post and stop sharing it?"

Most of the women were simply surprised. They had never realized that their comments were visible beyond where they were posted and had never really contemplated the person on the other side of the screen who would be reading what they wrote.

Caitlin no longer finds Internet memes mocking individuals humorous and calls out those who do. "Each one of those people is a real human being, a real person whose world imploded the day they found themselves to be a punch line on a giant stage," she wrote. "I know what it's like to be the person in that horrible photograph. I can't inflict such pain on someone else."

The Flouncer

"I can't live like this. It's too much."

—Jessica Valenti[16]

But not everyone can be so fearless as those who fight back. For many dealing with cyberharassment, a strong impulse is to go into digital hiding, either deleting or privatizing their social media accounts. So many people have publicly quit social media altogether that this move now has a name: the flounce. Feminist writer Jessica Valenti, a *Guardian* columnist and author of *Sex Object: A Memoir*, took to Twitter in July 2016 to say that she was through, after receiving rape and death threats targeted toward her five-year-old daughter.[17]

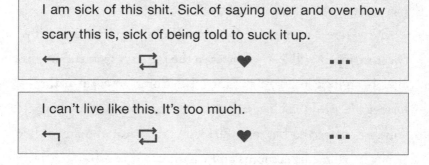

I am sick of this shit. Sick of saying over and over how scary this is, sick of being told to suck it up.

I can't live like this. It's too much.

Shrill: Notes from a Loud Woman author Lindy West also penned a piece for the *Guardian* in January 2017, claiming that she was abandoning Twitter after years of online abuse, mainly because of the company's lackluster response to policing its haters. "I have to conclude," she wrote, "after half a decade of troubleshooting, that it may simply be impossible to make this platform usable for anyone but trolls, robots, and dictators."[18]

Lifestyle video blogger Anna Saccone vocally yanked herself off-line, filming one last goodbye YouTube video titled "Why I Quit YouTube: Eating Disorder, Body Shaming" for her million-plus followers. She explained that fat-shaming comments like the #SaveObeseAnna hashtag campaign felt unendurable after her miscarriage caused her to be in an emotionally fragile state, and she also revealed that she had previously suffered from bulimia. "When things are going fine in life, when things are going all right, it's not that big of a deal, the criticism and the negative comments... You can kind of just brush them off, or at least I could," she told viewers. "But when something goes wrong... I feel like that, to me, made it a hundred times worse... I know it shouldn't matter, but it did... It did get to me, because of the mental state that I was in already."[19]

New York Times editor Jonathan Weisman wrote a goodbye tweet saying, "I will leave Twitter to the racists, the anti-Semites" after being targeted by hateful trolls.[20] Some go even further—former Harvard sex blogger and feminist activist Lena Chen left America, moved to Berlin in 2013, and deactivated many of her public social media accounts and websites.

Some pundits worry about the fact that so many people are being forced out of their industries, their jobs, and most of all, public discourse. "The cumulative effect of shame campaigns and intimidation strategies is that millions of people simply flee the field, leaving the battle to the most extreme voices or to those people who've slowly developed the thick skins necessary to maintain a public presence," David French writes in the *National Review*. "If most people believe that the price of engaging in the world of ideas is direct attacks on their children—or the potential loss of a job they love—then they'll simply bow out."[21]

If you need to take some time away from the Internet, that is understandable. "You have to balance your desire to speak up with your desire to take [care] of yourself," says Emily May, founder of Hollaback, which fights against street harassment and has now taken its crusade online. "There's a right time to speak up and a right time to prioritize yourself over the rest of the world."[22]

As privacy expert Theresa Payton says, a little "digital detox" never hurt anyone.[23]

The Humanizer

"The only thing you have in control is your attitude toward it."

—C. D. Hermelin[24]

C. D. Hermelin was living the dream—a broke-yet-aspiring freelance writer who'd recently moved to New York City—when he came up with what he thought would be a unique money-making schtick. Calling himself the Roving Typist, he'd sit with a typewriter in a public park and sell customized, "flash" fiction to passersby for a small fee. He tested out the concept in New York's Central Park and Washington Square Park, but he found his greatest response came from setting up a bench on the High Line, the one-and-a-half-mile-long aerial walkway along Manhattan's Lower West Side.

In the summer of 2012, a Brazilian tourist named Carla snapped a photo of him hunched over his typewriter and posted it on Reddit. The image, titled SPOTTED ON THE HIGH LINE, ended up on the site's front page, where millions viewed it. Unfortunately, the photo had just one problem: the shot cropped out his Stories While You Wait sign, so there was no reasonable explanation of exactly *why* C. D. was sitting in the park with his typewriter perched on his lap. Reddit's feisty posters responded with predictable animosity. "I have never wanted to fistfight someone so badly in my entire life," one commented. "It's because of these guys that bullying is so hard to stop," said another.[25]

"Everybody in the thread hated me," C. D. would later recall. "Hundreds and hundred and hundreds of people discussing why they hated me, and why I deserved to have my typewriter smashed over my head."[26] C. D. says reading all those comments made his heart "hurt."[27] The words brought back all the memories of being bullied and threats of violence that he had experienced in high school, and they even made him nervous to return to the park. "It only takes one person to act on their threat."[28]

But he decided to take the brave step of jumping into the Reddit thread to explain himself. "So many people were asking the question, 'Why is he doing that?' that I thought, 'Hey, why not answer?' I wasn't sure it was the right move, [but] if it turned out to not be, I could always delete my comment." Remarkably, the tone shifted. People deleted mean comments, apologized, and even began saying that what he did was pretty cool. "I think it had to do with timing and my actual explanation—a lot of the people on that site realized that it was something they'd like to try," he says. "I also think my tone had something to do with it, which I hoped came off as 'respectfully bemused.' If I had gone in and lectured on antibullying or something, I'm pretty sure I would not have had the same response... I found that being levelheaded and positive in the face of that nastiness worked."[29]

Still, this turned out to be the quiet before the storm.

Days later, the photo showed up again, this time poised to become an Internet meme, with lines of type inserted on the image that read: YOU'RE NOT A REAL HIPSTER UNTIL YOU TAKE YOUR TYPEWRITER TO THE PARK. This time, it went everywhere—pinned on Pinterest, posted on Tumblr— winning him plenty of hipster hate. C. D. describes his state of mind as "anxious confusion." He knew trying to get the image taken down was a lost cause. Despite telling himself not to, he would sit and scroll through all the hateful comments, numbing himself, in a sense, so that he could turn off his computer without wondering whether he had missed anything. "I talked to a lot of friends on the phone," he recalls. "It was good to hear the voices of people that knew me and loved me for me, who understood the absurdity of what was happening... I had to decide to believe that what was driving hate was the picture and the placement online, not me."[30]

Ultimately, he found himself responding yet again. A year later, he wrote an article for The Awl titled "I Am an Object of Internet Ridicule, Ask Me Anything," which he says helped him retrospectively sort out his "lingering emotions."[31] Today, he only wishes he had done so sooner. His quirky claim to fame proved to be a good conversation starter among fellow scribes at his MFA creative writing program at the New School, and it attracted the interest of a filmmaker who filmed a short documentary on him called *The Roving Typist*. "The reaction that it got was more than I ever could dream," he said of the response to his Awl piece. "It wasn't just about bullying online, it was sort of bigger than that, like 'What is our culture doing, what is the Internet doing, really?' It sparked these conversations... It was a huge moment for me... I definitely go online hoping that I see that my article has changed something a little bit—and I think it has."[32]

C. D. is now a literary agent helping others pursue their writing dreams while he writes his own novel, and he continues to sell his made-to-order stories, poems, and horoscopes at weddings and other celebrations. His best advice to how to handle a digital disaster? "Unfortunately, you are powerless to change what's happened. The only thing you have in control is your attitude toward it."[33]

The Empathizer

"I've become that person where it's hard to leave something alone. Someone will say, 'Just ignore it.' I can't."

—Carol Todd[34]

Is it possible that reaching out to your tormenter with a sympathetic ear can actually make a difference?

Carol Todd is a Canadian mother who took the tragic 2012 suicide of her daughter, cyberbullying victim Amanda Todd, as a call to action. She created the Amanda Todd Legacy Society and began its crusading work against cyberbullying and sextortion (read more about Amanda's story in chapter 9). Carol has put herself in the limelight, speaking at conferences and to organizations across North America. As a result, unbelievably, she herself has been subjected to a barrage of online harassment. But one message stuck out more than any others. About a year ago, Carol received a note in her Facebook inbox that read: You're an alcoholic whore and you should die just like your daughter.

Don't feed the troll.

In her head, Carol knew that's what she was supposed to do—nothing. But something made her ignore that sage advice. "I screen captured it and posted it on my Facebook wall so others could see," she told me in an interview.[35] Many of her online friends began offering words of support and criticizing the troll. Then two weeks later, she got another unexpected message—this time an apology: Miss Todd, I'm so sorry, can you ever accept my apology?

Carol was dubious that these words from the same troll could be sincere, so she simply ignored it. Then he sent another. Eventually, her curiosity provoked, she responded: Why did you do this?

His answer was disturbingly, sickeningly honest. "He said, 'Because it brings [me] joy.'" She laughs harshly, remembering this jarring response.

She demanded to know why this would bring him any joy: Do you actually know who I am?

He responded: No. I just know you lost a kid.

Carol was stunned to learn that the young man took pleasure in finding parents who had lost children and trolling them, often viscously. "He told me he was really horrific; he would tell parents they should go kill themselves," she recalls.

Perhaps it was the mother in her, but Carol listened, and the young man's story eventually came tumbling out. He was only twenty-one, raised by an alcoholic father and a mentally disturbed mother, and was clearly suffering from depression. He had been bullied himself and was homeschooled as a result.

Carol Todd is a bigger person than most of us. She bravely allowed this troubled young man into her life, offered him a sympathetic ear, and even opened her inbox to him nightly. "Every night, he messages me to ask how my day was," she says.

At her suggestion to just "go do random acts of kindness," he took a group of homeless people in his town to Subway and bought them all lunch. "He literally changed his behavior. It makes me feel good. He needed someone to talk to him and guide him."

It's hard to imagine that reaching out like this would be the right response to a troll, but others have found this to be true too. Lindy West shared a similar story of confronting her own Twitter troll on the radio show *This American Life*. Lindy had waded into the raging online debate between feminists and comedians about whether jokes about rape can ever be funny. In response, a cruel troll created an account with her dead dad's face and the name PawWestDonezo, a reference to the fact that her father, Paul West, had recently succumbed to prostate cancer. The bio read, "Embarrassed father of an idiot." Lindy published an article in response, saying how much those actions hurt—and why she would choose to respond to trolls. "I talk back because Internet trolls are *not*, in fact, monsters. They are human beings—and I don't believe that their attempts to dehumanize me can be counteracted by dehumanizing them. The only thing that fights dehumanization is increased humanization."[36]

To her shock, she received an apology by email from that troll, who wrote this explanation:

> I think my anger toward you stems from your happiness with
> your own being. It offended me because it served to highlight
> my unhappiness with my own self.[37]

Jen Royle, the embattled sports reporter, also received an

unexpected response from one abusive Twitter troll, after she pointed out, from reading his profile, that he was a father. She received (through her attorney) this email in response:

> You said some things back to me that really made me think. I AM a husband and a father, and that is NO WAY to be speaking on a public forum that at some point my son could see. I don't know you personally, and I don't know how you live your life, and it's none of my business. I don't expect you to accept my apology, since I'm sure you are very upset with me. I'm just doing the only thing I know how to show you that you were 100 percent right, I am a bad excuse for a human being, and I am deeply, deeply sorry for any distress I could have caused you with these posts. I'm sure you are a very nice person, and I'm sorry I got on the wrong side of you.
>
> Just please know that I regret everything I said and I am completely sorry for causing you any pain or stress. I also want you to know that this has caused me to take a step back and look at my life and how I'm living it. So thank you for calling me out and making me look at myself in this whole situation. It will definitely change how I do things and make me look at my decision-making process in the future. Once again, I am very sorry for the comments and any pain I have caused.

Surprising as it may seem, in some cases, reaching out with a compassionate ear could be exactly what some trolls secretly need.

The Apologizer

"I'm deeply ashamed of myself, and I'm truly, truly sorry."
—Mike Gallagher[38]

The Internet is the world's largest megaphone, especially when you're hosting a radio show. You shoulder the responsibility of knowing that people from all over the globe are potentially listening. When Mike Gallagher, of *The Mike Gallagher Show*, chatted with his guest, Fox News anchor Chris Wallace, live on the air, they made some distasteful comments about Kelly Clarkson.

What started off as a random conversation about pizza and Gallagher's weight quickly took a nosedive into fat-shaming the talented and beautiful Clarkson, with Gallagher saying, "Have you seen Kelly Clarkson?... Holy cow, did she blow up." Wallace, who, given his years in broadcasting, should have known better, followed up with, "She could stay off the deep-dish pizza for a little while."[39]

Their crude words swiftly went viral—as most online hate remarks do. Fans of Clarkson and listeners from around the world were outraged. People who were passionate about weight issues and body image, and others who were simply disgusted with how these two *adults* had demeaned another person, were emailing Gallagher and posting on social platforms including Facebook about this offensive behavior.

Gallagher issued an online apology almost immediately in a blog post:

Tubby Mike is the last person in the world who should bring up anyone's weight. I couldn't possibly feel any worse than I do for making an observation that led to the conclusion that I "fat-shamed" this talented and classy entertainer. It was a really stupid thing for me to do.

However, it was a few days later, on his next live radio show, that Gallagher truly became the *apologizer*.

Many of us could write out an apology in a statement, as Wallace did, especially after being pressured by our colleagues. Wallace also faced Internet backlash, and his former Fox colleague Greta Van Susteren called on him to apologize. He then released this formal statement: "I sincerely apologize to Kelly Clarkson for my offensive comment. I admire her remarkable talent and that should have been the focus of any discussion about her."[40]

But it was Gallagher who took to the airwaves, where the shaming originated. His voice expressed a tone of genuine apology, noting that he had spent the weekend reading the harsh emails and comments from everyone and that he understood that he deserved them, "*a thousand percent.*" He said, sincerely, that it was extremely "insensitive and thoughtless" of him—adding, "I had no intention of being mean-spirited, [but] that's exactly how it came out, and I'm just deeply, deeply sorry."[41]

Isn't this how cruelty sometime happens? When we don't think or *pause* before we speak or post?

Gallagher ended his verbal apology by recognizing that the experience was a "turning point" for him. "The fact is, my words and the words that I use in describing people that I disagree with can be very painful and can be hurtful, and I don't spend enough time recognizing this." He acknowledged that there is never a reason to "attack somebody personally" or to "tear people down," online or off.

Would his fans and listeners forgive him? Since this 2015 incident, he received his due backlash, but today is ranked as one of the most listened to radio hosts in America.[42]

And how did Kelly Clarkson feel about all this?

Clarkson didn't respond to Gallagher and Wallace's remarks, but she appeared on *Ellen* shortly after the incident and spoke generally about the negative comments she'd been receiving about her weight.

"I think what hurts my feelings for people is when I'll have a meet-and-greet after the show, and a girl who's, like, bigger than me will be in the meet-and-greet and be like, 'Oh God! If they think *you're* big, I must be so fat to them,'" she told DeGeneres. "It's like, you're just who you are. We are who we are—whatever size."[43]

Being the target of hurtful comments isn't easy for anyone, and it affects everyone differently. Whether you are an activist for change online, or someone who finds the human side in today's digital space, or even someone who had an oops moment that you were big enough to apologize for, it's important to know you're never alone!

When I first experienced trolls and cyberharassment, there weren't a lot of outside resources to go to for support. I didn't have the tools that are available to us today, nor did I know the steps to take when you are being attacked. Thankfully, today, that has changed, so I'd like to share these resources with you now.

Let's go.

CHAPTER 8

SEEKING OUTSIDE HELP

"Freedom of speech online doesn't mean anything if people are not free from abuse and harassment."

—HeartMob[1]

If your attackers are coming after you hard, it might be time for a more forceful response. Has your online reputation suffered irreparable damage? Has it gotten so bad that you are fearful for your safety? Should you consult an attorney or file a police report?

It's certainly possible that you could be facing some of the mentally unstable people on the Internet who take trolling to the extreme. One woman, "Sarah," first got sucked into an online flame war with one of these people years ago, when she was in grad school and brushed with an anonymous poster on Lena

Chen's *Sex and the Ivy* blog. "At that point, I'd never engaged with anyone online," she recalled when I spoke to her. "I made a couple of comments. It didn't even occur to me anyone would care; I was nobody."[2] But her harasser, in retaliation, targeted her for years, creating rambling posts calling her a fat-ass, questioning her professional work, and even claiming that she had rape fantasies.

In fact, she says, this deranged stranger ultimately went after a dozen women who posted comments supporting her, trashing their reputations and getting at least one fired from her job as a teacher. "I've never encountered a situation quite like this one," Sarah admits. "I was being called a fat, ugly slut who wants to be raped. He spent eight years, on and off, trying to make me unemployable, undateable, trying to make me a target for crime—he's accused me of having STDs, publicly claimed I was fired from jobs for sexual misconduct. That level of effort into trying to hurt someone... It is hard to understand. All I can think is this is someone deeply unhappy, and probably a sociopath. There are people out there who will think nothing of trying to hurt other people."

"This is not trolling," states researcher Lindsay Blackwell, in her talk titled "Trolls, Trouble, and Telling the Difference."[3] "This is harassment. It is violence, and it is very, very difficult to control." According to a 2014 Pew Research Center report, 18 percent, or nearly one in five, of Internet users surveyed were the victims of "severe" online abuse, such as stalking or physical threats.[4] Advising victims to get off the Internet doesn't work either, Blackwell adds. "The Internet is real life. For many of us, it's where we make a living, it's where we make friends, it's where we live our lives, it's an extension of where we live our lives... Telling victims of online harassment to log off...[isn't] helpful... It's so, so important that we stop telling victims of harassment to not feed the trolls."

Sarah's experience has impacted her life in so many ways. When it came to dating, she would only give out her first name, then field comments from suitors who came across what was posted about her online. "I had someone tell me I shouldn't mention it to men, [that] they would think I invited drama into my life," she recalls. Now an MBA student seeking employment, Sarah has needed to discuss her situation with career counselors, add explanations to her cover letters, and warn potential roommates. She has filed police reports and consulted with two pro bono attorneys, but to date, has been unsuccessful in unmasking the identity of the perpetrator, leaving her understandably bitter. "You can't sue someone if you don't know who they are," she explains. She hesitates, then adds flatly, "This sounds awful, [but] I would love to destroy his life in the way he's tried to destroy mine."

Controlling a Disaster

Reading comments online, especially twisted truths or outright lies about yourself, can be horrifying—I know this firsthand. The emotional toll that it can take on a person is enormous. Know that you don't have to go through this alone. There are many outlets to help you through this cybertorture.

What options are available for the average Joe, who doesn't

have an entourage, a high-powered publicist, or the star power of a celebrity to mobilize support or fund a legal battle?

To take control of a digital disaster, begin with these basic steps:

+ **Document the attacks.** Take screenshots of all the evidence. You might want to just push delete, delete, delete. But if things escalate, you'll need to have some documentation. Print it out, keep it in an online folder, put it on a thumb drive, download any videos to an external hard drive—but do save it. Some even advise using a web-archiving service, such as Page Vault, which officially documents the date, time, and web address, allowing it to be legally permissible in court. This is an area where your friends and family can help you. Ask them to monitor the abusive content for you, so you don't have to read it over and over again. "There was a point where I started to have an anxiety attack every time I thought about Googling my name; I felt like I needed to see if anything new was posted," recalls Sarah, who had her father take over that unpleasant task. "Mitigating how much I was exposed [to] was really important."

+ **Block the offenders.** Blocking functionality is available on social media platforms, as well as phone calls, texts, apps, and email. One tool offered by Twitter, under the guise of protecting your well-being, is quality filtering, which prevents you from seeing anything that a specified poster has posted about you. In November 2016, Twitter also expanded this mute feature to include specific words or users you choose to block.[5] But some experts are not fans of this "ignorance is bliss" credo. Blackwell points out that if you enable this feature, there is no way to be aware of—and stop—what is being said behind your back.

+ **Report the offenders.** Review the website's or platform's Terms of Service (TOS) or Code of Conduct, to identify what actions are considered violations, then politely ask the service to remove offensive comments, in accordance with its guidelines, and to ban the violator from the platform. Beware—some sites, especially those that seem to foster harassment and revenge porn, have been known to thumb their noses at victims and reprint emotional takedown requests, so don't get overwrought in your tone. Stick to boilerplate legalese.

+ **Try to identify the attackers.** Are you being harassed or stalked, and it's escalating? Maybe you are fed up with the cyberslime an anonymous user is posting about you. To identify that person's IP address, you will need to file a crime report with law enforcement, says California Senior Officer Mike Bires. "After reviewing the facts of the case, the detectives can obtain a warrant, which can be served upon the social media platform in question, requesting all of the information relevant to the case and the suspect."[6] But if you don't want to contact the authorities, there are other ways to find out who's behind the IP, suggests security expert Theresa Payton, CEO of

Fortalice Solutions and coauthor of the book *Protecting Your Internet Identity: Are You Naked Online?*. "When you are in the digital world, you can feel powerless," she says. "Consider hiring a cybersecurity company to unmask the aggressor, if the case is more serious. Unfortunately, my company has been increasingly hired to do this. It's not a sure thing that you can unmask them, but often you can, if [you have] suspicions [about] who is harassing [you]. Everyone has a digital pattern, so watching the harasser's patterns [may lead] you to the real-life person."[7]

‧   **Cut the criminals off.** If you ever find yourself being extorted for money over explicit materials, treat it like you would any other form of blackmail, recommends ReputationDefender's Rich Matta. "Cut off all ties with the extortionists. Block their email addresses and social accounts. Realize that a payoff is unlikely to change their behavior or resolve the issue. If videos or materials have been posted 'privately' along with a threat to go public, fill out the appropriate online forms to request removal from YouTube, Vimeo, Google, GoDaddy, or whichever hosts or service providers are hosting the explicit material. If appropriate, contact the authorities."[8]

Officer Bires says, "Agencies throughout America are receiving information, tips, bulletins, and training every single day on social media and cybercrimes." He recommends that if you become a victim of sextortion, report the crime to your local law enforcement. Bires continues,

"Granted, not every police department has the expertise to investigate such technical crimes, [but] these same agencies know there are law enforcement professionals who can assist them in completing an investigation."[9]

How Are They Even Allowed to Post That?

Many victims wonder why there are no legal repercussions for the many websites or platforms on which these online attacks are published. The reality is that small websites and large behemoths like Google and Facebook alike are protected from being held accountable for the content others post on their sites. Known as Section 230 of the Communications Decency Act, the law places the legal liability on the *poster* of the information, rather than on the platform or host itself. In practice, that means that you need to go after the person who posted the abuse—and for that, you'll often need a court-ordered subpoena forcing the website to identify the source.

Still, there are a small handful of scenarios where you will easily be able to get the content deindexed (taken out of search results on engines like Google, Yahoo, and Bing, which reduces the likelihood of anyone seeing it) or removed outright from the original website. These scenarios include the following:

+ A disclosure of sensitive personal information connected with your name, such as bank account or credit card numbers, social security numbers, handwritten signatures, and other information that could lead to identity theft

+ A violation of the Right to Be Forgotten law (if you live in an area governed by the European Union or in Argentina)
+ An incident of nonconsensual porn

Since 2015, most online platforms, from Vimeo to WordPress, have updated their policies and will now remove any instances of revenge porn or nonconsensual pornography. The Cyber Civil Rights Initiative (cybercivilrights.org) has created an excellent "Online Removal Guide," which offers step-by-step instructions for requesting unauthorized images or videos to be removed from sites such as Facebook, Instagram, Twitter, Reddit, Tumblr, Yahoo, Google, and Bing.[10] Without My Consent (withoutmyconsent.org) also has a similar guide to getting nonconsensual pornography removed, called "Something Can Be Done! Guide," and is in the process of developing a state-by-state guide to various legal recourses in each state.[11]

+ Copyright infringement. You automatically own the copyright to photos that you take, including selfies. If someone has posted a photo that you took of yourself, without your consent, then that person has violated your copyright to that image. In this situation, you don't have to go through the effort of trying to prove a case of defamation, emotional distress, or invasion of privacy. As the copyright owner, you can submit what is known as a Digital Millennium Copyright Act (DMCA) takedown request, and most websites and platforms will immediately comply, or risk facing large fines. Applying with the U.S. Copyright Office for copyright ownership can give you an extra edge, but is not strictly necessary, and usually requires sending in copies of the photos as documentation, something that most victims of revenge porn are loathe to do. Several services will, for a fee, help you with this Herculean task. Three services that I recommend are: DMCA Defender, CopyByte, and (to do it yourself) DMCA.com. They will charge a set fee for removal of each image, and will waive this fee for minors. (But beware—if you are waging a battle with the person who has posted your photos, filing a DMCA takedown creates a formal complaint and a paper trail that will include personal details about you, such as your home address, that will be viewable by your harasser.)[12]

Suing Them

What if the site or platform refuses to comply? Many rogue sites have a policy of refusing all pleas, and some outlets, like traditional newspapers, have a policy of not removing content on journalistic grounds. Sometimes, if the damage to your reputation is severe, consulting a lawyer about a civil case is the best way to go.

An attorney can begin with a simple cease-and-desist letter, which puts your harasser on notice. "A properly worded cease-and-desist letter will often stop the troll in his or her tracks," says attorney Mitch Jackson. "It lets the troll know that you've retained counsel and are taking the comments seriously. Because the letter explains the legal action the troll is exposing himself to (lawyer fees, court costs, and financial damages), a certain reality check takes place."[13] Attorney Christina Gagnier, of Without My Consent, says an attorney can also file for a civil harassment restraining order. These restraining orders are different from ones that require an abusive ex to stay away from you—they can be used to prevent a defendant from doing or even publishing specific things.

If you decide to pursue legal action, there are several potential avenues, depending on your state's laws, for a lawsuit: defamation, invasion of privacy, intentional infliction of emotional distress, stalking, and harassment. Your attorney can also explore causes of action that would apply in federal court, such as copyright infringement, federal stalking claims, and computer fraud and abuse claims.

According to Jackson, the reasons to file a lawsuit are to: (1) identify the troll, (2) stop the harm, (3) obtain an injunction, and (4) seek monetary damages. "By filing a lawsuit, your lawyer will have subpoena power and access to the social media platform's records as they pertain to the troll," he says. "It brings in a judge who will oversee the entire matter and make sure the troll follows his or her rulings and orders. It also sends a message that he or she is going to be held accountable and that his or her life is about to change, for the worse… If the criminal or recreational troll causes you physical or emotional harm, or financial damages, then the civil court will allow you to be made whole [by awarding monetary] damages."

Of course, pursuing a legal case is no easy task. To meet the standard of defamation, for example, your case must meet two thresholds: it must be provably untrue, not just an opinion, and it must have financially quantifiable damages. "The challenge is finding and determining the identity of the wrongdoers, and then holding them accountable," says Jackson. "Often, the person who caused the harm doesn't have the ability to pay a judgment, so the victim ends up out-of-pocket with a civil judgment that isn't worth any more than the paper it's printed on."

Maybe you're fearful that dragging your case through the legal system will bring fresh attention to your plight, a phenomenon known as the Streisand Effect (named after the 2003 incident in which the famous singer tried to have images of her house removed from a collection of photos archived by the California Coastal Records Project, on the grounds of invasion of privacy, a strategy that backfired by bringing her increased scrutiny and spreading the images widely across the Internet).[14] If so, you and your lawyer can explore whether you can file suit without publicly disclosing your identity, a tactic known as filing pseudonymously, as a Jane or John Doe. But as Jackson warns, "The reality is that the true name of a victim will almost always come out at some point, and so it's important for victims to know the pros and cons of using this approach."[15]

Cybersquatting is a common way people can infringe or misuse or abuse your good name online, especially if you are well branded. Another little-known technique for fighting back against online defamation or cybersquatting is obtaining trademark protection for your name. Let's say you're taking on a website called yournamesucks.biz. You'll have a much stronger case to get that website taken down if you can claim that the site is infringing on your trademark, according to attorney Bradley Shear, who practices social media privacy law in Bethesda, Maryland. The U.S. Patent and Trademark Office charges between $225 and $400 to file for a trademark, which is easier if you have a distinct, widely recognized name and can argue that your name or business name is associated with a particular good or service (think Kleenex and Band-Aid).[16] "One way of protecting your organization or yourself is to go through that process," Shear says. "In the long run, it's a lot cheaper than buying every [possible] URL."[17]

If you want to speak to a lawyer but cost is a concern, check with your local law school or state bar association to see if they might offer any pro bono services. Several law schools are now offering low-cost or free representation for victims of cyberharassment, such as the Tyler Clementi Institute for CyberSafety at New York Law School and the Intellectual Property, Arts, and Technology Clinic at the University of California–Irvine School of Law.

Even with legal aid, this can be a long, expensive journey, and in many cases, online attacks, while horrible on a personal level, don't violate the law. "From a legal perspective, even if you think someone is bad-mouthing you online, it is protected opinion," says Shear. "When it comes to defamation, slander, and libel, every state has their own definition. If you want to go that route, that's very tough, and if someone is a public figure, there [are] heightened protections."

Sadly, even with an attorney on your side, it can be all but impossible to stop something from spreading once it is out on the Internet. "I talk about what the law can and can't do," says Shear. "It's a challenge to put the genie back in the bottle. You can try all sorts of maneuvers, and someone can [still] post it again. You're playing legal Whac-A-Mole."

Pressing Criminal Charges

If physical threats have been made, your personal address or other information has been published in a threatening manner, or things have simply escalated to where you are fearful for your safety, you need to contact law authorities. Even if your state doesn't have specific anti-cyberharassment laws, those already on the books, such as stalking or harassment laws, can apply.

A 2014 online poll found that 29 percent of people experiencing online harassment said they were scared for their lives.[18] In 2013, feminist activist Caroline Criado-Perez became the recipient of such threats because of her campaign to honor writer Jane Austen on the British banknote. She shut down her Twitter account and ultimately engaged the police to arrest and prosecute the worst offenders, which is an option under British law. Sadly, as events like the Gamergate saga of 2014–15 showed us, online death threats are not always so easily or successfully prosecuted in the United States. After expressing vocal opinions about sexism in video games, respected industry professionals such as Brianna Wu and Anita Sarkeesian became real-life targets, receiving dozens of death threats until they felt compelled to flee their homes or cancel public appearances. Yet federal prosecu-tors have declined to charge multiple suspects with any crimes, despite, in some cases, the FBI's having obtained confessions on video, sparking a new wave of anger and protest from victims.[19]

One common intimidation tactic that criminal trolls use is known as doxxing, publishing your home address and phone number to make you feel panicked or under the threat of real physical danger. If this happens to you, as one doxxing guide advises, don't give your attackers the satisfaction of knowing that their information is correct or that it has fazed you. Instead, make no response at all, to send the message that the attack was a "failure."[20]

Another growing harassment tactic, once your home address has been made public, is swatting. This truly terrifying scenario is when someone makes an anonymous, false emergency call report-ing a shooting in progress to local law enforcement, resulting in a SWAT team descending unannounced upon your home, guns drawn. Although this may not seem to fall under the heading of a digital attack, it is often the endgame of those who are doxxing their victims. While these vicious attacks are rare (the FBI estimates that four hundred cases of swatting occur annually[21]), they are impossible to prevent. While there are federal laws against other forms of false alarms, such as calling in a fake bomb threat or terrorist attack, there is nothing currently criminalizing the act of swatting—although federal legislation is in the works to change that.

If you need to report online harassment, don't stroll into your small-town, local police station, which may not have the technical know-how to handle the case. Instead, try your county or state police, who may have more resources.

Nationally, the FBI has an online Internet Crime Complaint Center, but it generally has bigger fish to fry. In a *Mother Jones* profile, Congresswoman Katherine Clark relates a story of meeting with the FBI on behalf of one of her constituents, Brianna Wu, the high-profile Gamergate target, and basically being told that the agency was not interested in the issue.[22] "Frankly, the FBI told us cases of online abuse were not a priority," Clark says.[23]

Even though it may be a challenge, find a police department that may have time on its hands for your case. "Don't get discouraged if they're not hyperresponsive," says attorney Gagnier. "Sometimes, in law enforcement, the person at the front desk, they don't know this is even a crime. They're not trying to be dismissive, they just don't get what you're talking about."[24] Remember, it is up to the discretion of law enforcement and the district attorney whether or not to file a case. Bringing along a binder with all your proof printed out will help.

Officer Mike Bires, who founded the blog *LawEnforcement .Social*, insists that you must be your own best advocate. "'I went to the cops and they didn't do anything' is a statement heard probably a thousand times a day in this country," he says. "What those who say this conveniently leave out is the part of the story which causes law enforcement 'not to do anything.' The majority of these challenges are created by the victims. They may not retain any proof of the harassment—things like not having kept any records, proof, or evidence to assist with their investigation—and when it amounts to a repetitive, continual problem, they have little to show law enforcement."[25] Another forensic challenge, he says, is "proving" who posted an explicit photo or video of the victim if the poster's identity is concealed. Once you get the weight of law enforcement behind you, however, police will be able to forensically preserve the evidence and then help get the offensive material removed.

"Burying" It

If your main concern is the damage done to your online persona, one option is hiring an online reputation management (ORM) firm to clean up your cybermess.

Online reputation management is estimated to be a $5 billion industry made up of hundreds of firms, according to the 2013 *New York* Magazine piece, "Scrubbed."[26] Industry leader and pioneer ReputationDefender, sister company of Reputation .com, was founded in 2006 by Michael Fertik, author of *The Reputation Economy*. Since then, many competing—and some less scrupulous—firms have popped up, so be sure to select one carefully. All reputation management agencies basically work on the principle that you typically can't remove negative search engine results—but you can front load your page results with positive links, from LinkedIn profiles to glowing news articles to a Google AdWords campaign.

Why does this work? Simple. The vast majority of people Googling someone *never scroll past the first page of results.* "Studies have consistently shown that 90 to 95 percent of people never go past page one of search results, and about 99 percent never go past page two," says ReputationDefender CEO Rich Matta. "If you control the top of your search results, you are effectively removing the negative stories [that] people actually see about you online."[27]

DO THE DEAD HAVE A RIGHT TO PRIVACY?

When eighteen-year-old Nikki Catsouras was tragically killed in an auto accident, her bloody, lifeless body was left in the most horrific circumstances—nearly decapitated. After disturbing images from the scene were disseminated by two dispatchers with the California Highway Patrol, it wasn't long before they spread throughout the cyberworld. Trolls began using the images to taunt her grieving parents, who turned to Michael Fertik at ReputationDefender for help. The renowned ORM firm worked diligently and was successful in persuading about twenty-five hundred websites to take down the photos. Even so, Fertik told *Newsweek*, "Long story short, it became a virtually unwinnable battle," as the images continued to spread to new sites.[28] Legally, the family launched a battle pitting free speech against the privacy rights of the deceased. After four-and-a-half years, the resulting lawsuit altered California state law, which now offers privacy rights to surviving family members when it comes to mortal images.[29] "The fact is that we will never get rid of the photos anyway," says Lesli, Nikki's mother. "So we have made a decision to make something good come out of this horrible bad."[30]

Another online reputation management firm, BrandYourself, was founded with the mission of spreading affordable reputation repair to the masses. Patrick Ambron and his Syracuse University college roommate, Pete Kistler, formed the Lancaster, Pennsylvania–based business in 2009 after they noticed that Kistler, a new grad, kept getting passed over for jobs, despite a stellar résumé. Finally, he discovered the underlying issue: he shared a name with a convicted drug dealer.

BrandYourself's do-it-yourself program allows you to screen your results, from Google searches to your own social media postings, and flag possibly offensive comments. (Those who need more extensive damage control can also sign on for paid services, or even hire a whole team to help them.) Famed for turning down an investment offer on TV's *Shark Tank*, the company now has half a million users. Horror stories about Ambron's clients easily rattle off his tongue, like the man who started a new management job at a financial services firm the week they were busted for criminal activity, and whose name was forever associated with press clippings of the scandal. "For a lot of people," he says, "it's unluckiness—you date the wrong person, your friend tags the wrong photo."[31]

How does BrandYourself go about helping its clients reinvent their image? "It's not rocket science," says Ambron. "Keep yourself clean, monitor it, if someone trashes you, try to get better stuff [about you] above [those negative results]." Here are some of its creative strategies to stack up positive search results:[32]

- Put out a positive press release on PR Newswire. Yes, anyone can write a press release about some positive life event, such as a job achievement, founding a website, etc., and put the news out via the newswire.
- Comment on news articles. A thumbs-up on a comment on a credible news site such as NYTimes.com can help, as well as a link back to your own web page.
- Create profiles or comments on credible sites with .gov or .edu domains, such as your local library or government websites.
- Write a blog post for a site such as the Huffington Post or Patch.

All this may seem like a savvy way to run around search engines, but Ambron insists that what his firm does is not manipulation or black-hat tricks. "We're not trying to outmaneuver Google," he contends. "We're empowering people to build relevant presences so they don't get screwed up and blindsided by irrelevant things that suddenly define them."

Finding Online Support

A decade ago, there were few resources for those suffering online harassment, but today, there are many groups providing help to victims of digital attacks, from harnessing simple messages of support to full-on legal aid.

When Olympic gymnast Gabby Douglas received hate tweets during the 2016 Olympics over her hair and even her lack of a smile, a Twitter campaign—started by Leslie Jones—sprung up with the hashtag #Love4GabbyUSA, to rally support from a herd of fellow celebs. Everyone from Shonda Rhimes to Monica Lewinsky jumped in to bolster the gold medalist.[33]

Wouldn't it be nice if you had a high-profile friend who could do the same for you?

Enter **HeartMob**. The online platform (iheartmob.org) was started by Hollaback so that individuals being harassed online can create a report and, within an instant, rally teams of people who send messages like "Stay strong!," "I'm with you," "We have your back," and "You are not alone!"

Kate Bigam, the young Ohio woman targeted by anti-Semitic trolls, turned to HeartMob for support on a friend's recommendation. "When something like this happens, it can be difficult to remember that this hateful minority is just that—a minority. You start to feel really alone and to doubt yourself. You know they're just trolls, but it's scary and it's hurtful and it still impacts your self-esteem. To have all these people chime in and tell you they've got your back is a really powerful reminder that all these bad people in the world are evened out, hopefully, by so many good ones."[34] As Emily May, founder and executive director of Hollaback, a group that has been taking on street harassment for more than a decade, says, "The power of harassers to organize quickly and effectively to take down anyone that stands in their path could and should be countered by the power of the rest of us. There's something powerful about strangers ripping you to shreds, [but] the opposite is [also] true—there is something about complete strangers having your back."[35]

Need some "pest control"? Female journalists or writers can report their online harassment to **TrollBusters** (troll-busters. com), an online group started at the 2015 International Women's Media Foundation hackathon to provide supportive messages to those reporting an attack.[36] "We then send a warning into their social media feed that notifies trolls that we are monitoring and watching the stream," founder Michelle Ferrier, a former newspaper columnist turned Ohio University journalism professor, told *EContent* magazine. "Sometimes that's all it takes to get the harassing behavior to end."[37] TrollBusters's next project: studying the Twitter feeds of female journalists, to analyze how the web of harassers operates and to develop ways to combat trolling.[38]

The Cybersmile Foundation (cybersmile.org), an antibullying nonprofit founded in 2010, offers Total Access Support, where trained advisers give guidance to those being cyberharassed. "We help thousands of people each year," says cofounder Dan Raisbeck. "Our trained advisor teams work around the clock answering inquiries from all over the world. Our advisors are mostly made up of volunteers who want to help others or offer their time in some way."[39]

Online SOS (onlinesosnetwork.org) is a California start-up that provides free legal referrals and counseling services to people suffering online harassment. Cofounder Samantha Silverberg, a licensed clinician, began it in 2016 with a pilot program helping thirty victims, and she aims to have a therapist and a legal resource established in all fifty states.

Crisis Text Line is a nonprofit that provides free, twenty-four-hour therapeutic assistance for victims in need of support, handling issues from depression to bullying (if you're in crisis, text START to 741741).[40] In March 2017, it was announced that Crisis Text Line was joining forces with Facebook. "We want to be wherever people are in crisis," said Nancy Lublin, Crisis Text Line CEO, in a prepared statement. "And we'll continue to be on the leading edge of technology, supporting people everywhere they are."[41]

Another resource to consider if an explicit photo of a minor is involved is the **National Center for Missing & Exploited Children** (missingkids.org). While it is best known for its crime-fighting and advocacy work to find snatched kids, the organization also fights against sexual exploitation and child pornography. If you're fearful that someone has extorted your son or daughter into sending them compromising images, call its CyberTipline (800-THE-LOST) to report a case of online sexual exploitation.

No one knows online abuse better than *Depression Quest* game developer Zoë Quinn, another high-profile victim of the Gamergate saga. She and Alex Lifschitz founded **Crash Override Network** (crashoverridenetwork.com) as a resource and help center for victims of online harassment. Funded by the nonprofit Feminist Frequency, Crash Override leverages its relationships with tech firms to help victims better navigate the reporting system. Since launching in January 2015, they have helped hundreds of people. These two wear the battle scars of wisdom: Gamergate was essentially set off when Quinn's exboyfriend posted a vengeful, nine-thousand-word screed against her in 2014—the computer folder where she documented her hate mail and threats exploded tenfold, from sixteen megabytes to sixteen gigabytes. "These aren't trolls," she told *Boston* Magazine in 2015. "And it's not online bullying. Bullying is something that gets you a pink slip in high school. These are people stalking, sending death threats, trying to get the cops to raid homes. These are criminals."[42]

Now that we have seen the worst of online hate and cyberhumiliation, it's time to discover the hope of people who have shown that there is life beyond shaming.

PART THREE

BEYOND THE
SHAMING

CHAPTER 9

REBOUNDING WITH PURPOSE

STORIES OF HOPE

"People cut other people down for entertainment, amuse-ment, out of jealousy, because of something broken inside them. Or for no reason at all...

    But the only way they win is if your tears turn to stone and make you bitter like them."

—Taylor Swift[1]

Many have turned their shaming into something we can all celebrate. Each one of these survivors is making a difference in lives today and teaching others from their experiences. Some have changed laws nationwide; others are cultivating kindness on campus. Let's take a look at these exceptional stories of triumph.

Cyber Civil Rights Initiative

"I don't Google my name anymore. Every time you see it,
no matter how long, it feels like the first time."

—Holly Jacobs[2]

In 2009, Holly Jacobs, a Florida student pursuing her PhD, found herself a victim of revenge porn after her nude photos were posted online and spread to hundreds of porn websites and Facebook profiles. "My life went from normal to just destruction and trauma," she recalled in an interview, telling me how the photos were viewed by her professor, HR department, and even her parents.[3] She went to the local police department and even the FBI, who told her that there were no criminal laws to cover her situation and that she should contact an attorney about a civil suit. She approached attorneys, bringing along a binder documenting her case, with strategically placed Post-it notes to cover images of her naked body. But for a poor graduate student, the cost to pursue a case was prohibitive.

Increasingly depressed, Holly began seeing a therapist, who diagnosed her with PTSD. Eventually, she got angry. "I reached a point where I went, 'This is not right; there's no pathway to justice for victims like myself.'" In 2012, she founded the website EndRevengePorn.org, a campaign that ultimately evolved into the Cyber Civil Rights Initiative (CCRI).

Since then, CCRI has rebranded to focus on nonconsensual pornography, to cover all situations where people have nude

photos of themselves distributed against their will. The organization is educating the public and providing victims with support
and legal referrals. Some six thousand emails from victims have
poured in to date, and a hotline that CCRI introduced in October
2014 now receives one hundred calls a month. The group has
helped persuade tech companies like Facebook and Google that
nonconsensual porn is grounds for instant removal and delink-
ing upon request. Most impressively, their campaign has created
sweeping new legislation nationwide. "When we started, there
were three states that had criminal laws [against nonconsensual
porn]: New Jersey, Alaska, and Texas. Now we're up to thirty-four
and DC in four years," says Jacobs. (For an up-to-date list, check
CCRI's website, cybercivilrights.org.) Additionally, a proposed
federal law has been introduced to Congress by California representative Jackie Speier.

Most importantly, Holly has noted a shift in public opinion.
"When I first started speaking out about revenge porn, it was
90 percent negative comments and slut shaming, and 10 percent
supportive," she says. "Now, it's 75 percent positive—it's really
come a long way."

The UnSlut Project

*"If you've discovered confidence in your own skills and…
your own power…you're going to be a lot less affected by
others' opinions of you."*

—Emily Lindin[4]

How is it possible to be called a slut when you haven't even had sex yet? No one knows the answer better than Emily Lindin, founder of The UnSlut Project.

Emily, a Harvard graduate who completed her PhD at UC Santa Barbara, didn't think much anymore about the painful years back in middle school, when her explorations with her sixth-grade boyfriend got her labeled the school slut. But her diaries hadn't forgotten one detail. On a visit home to her parents' house, she pulled them from the shelf of her childhood bedroom, where they were enshrined like a time capsule. There, she relived every high and low from 1999 to 2003: the crushes and the betrayals. The time someone on the school bus announced, "Who thinks Emily Lindin is a slut?" and every hand went up. The best friend who betrayed her and created that AOL Instant Messenger screen name DieEmilyLindin. It was the earliest days of the Internet, and she and her classmates were still exploring it as a fun medium. When those words on the screen turned against her, "It was a new type of affront, so surprising and shocking," she recalled when I spoke to her.[5]

Back then, there were no major social media sites where any slut-shaming that she was experiencing would have been seen by her parents and the wider community. But for today's teen girls, the outcome is often very different. "The Internet adds a whole level of anxiety to being attacked," Emily reflects. "If there had been an added layer of social media [for me], it would have been harder. It would have been inescapable."

Emily had considered suicide and even began cutting

herself at that tender age. So later, as an adult, she found herself feeling especially touched when she heard stories of cases like the seventeen-year-old Canadian girl, Rehtaeh Parsons, who took her own life after being gang-raped by classmates and then cyberharassed with intense sexual bullying. Stirred to action, in 2013, Emily began publishing her diaries first on Tumblr, then on Wattpad, hoping to give adults an "unedited, authentic look inside of a girl's brain…so they might understand why a girl might turn to self-harm." The Wattpad diaries went on to be viewed by more than 792,000 readers, and after articles ran in Jezebel and *Slate*, the project quickly took off within the feminist blogosphere, helping other victims of sexual bullying recognize that they are not alone. The UnSlut Project ultimately evolved to include a traditionally published book, a documentary film, and an online community for sharing stories. Today, hundreds of girls have published their own contemporary accounts of slut shaming.

Here are some of their stories:[6]

Someone saw me [masturbating] through my room window and posted it on social media without me knowing until people started talking about it around school… People at school began finding out about it one way or another, and began bullying me about it… It was a nightmare… Others who'd seen the footage, often times went around school whenever they could, and showing classmates/friends the inappropriate content on their phone like it was the latest episode of some popular show... As time went on, I began living in fear of being recognized in public, or potentially harmed everywhere I went. I rarely tried to go out anymore and did my best to keep my head down at all times in public. I had become ashamed of my existence, and depressed. —Anonymous

I was in 8th/9th grade and had a boyfriend. I was with him for about two years and lost my virginity to him. He asked for pictures of me naked, and I wanted him to like me a lot so I did it... He ended up showing all of his friends my pictures... I was branded a slut... I felt alone, I felt as if something was wrong with me. I swallowed a bottle full of pills, I ended up having suicidal thoughts frequently. —Ashley, Houston, Texas

After scrolling through the words of Denelle and Lola and Jessenia and Kate and Caitlin and countless others signed simply "Anonymous," most readers are left with sorrow and outrage. Why, in 2017, are we still putting girls as young as twelve and thirteen into one of two categories: prude or slut?

The UnSlut platform continues to comfort, educate, and be that safe space for women throughout the world who struggle with sexual harassment and bullying. Ashley from Houston finishes her own story with a message for others that I want to reiterate:

I'm now 21, happy and full of confidence... It's mind boggling that my story is one in thousands of similar stories. It shouldn't even be a normal thing that girls/women go through. I want girls to know that there are people out there that have gone through this and overcome it. Change can happen and it can get better.

Tyler Clementi Foundation

"We must stop this epidemic of cruelty, bullying, and humiliation! We need to do a better job at creating safe, respectful, and kind environments for our youth."

—Jane Clementi[7]

The Tyler Clementi Foundation was formed shortly after the death of college student Tyler Clementi. Although tragedies can bring positive outcomes, there will always be the emptiness of where Tyler could be today.

Tyler, at the age of eighteen, was a victim of the worst type of invasion of privacy, cyberharassment, and digital humiliation. In August 2010, Tyler started his freshman year at Rutgers University, excited about having the freedom to live openly as a gay man, as well as playing his violin at a high level of expertise at the university. Sadly, this all came to a screeching halt just weeks after school started.

One evening, Tyler asked his roommate, Dharun Ravi, for some privacy, because he had a date. Ravi agreed, but what happened next was horrific. Ravi secretly recorded Tyler and his date on his webcam and invited others to view it online. No one stopped this invasion of privacy. By the time Tyler discovered what had happened, it had already gone viral on Ravi's Twitter feed, and Tyler was the topic of ridicule and cyberhumiliation. He also found out that Ravi was planning to do it again. It was simply too much. Tyler decided to end his life by jumping off the George Washington Bridge a few days later.

To honor Tyler, his parents, Jane and Joe, organized the Tyler Clementi Foundation, a New York City–based nonprofit dedicated to fighting cyberbullying and helping countless youths, adults, educators, and communities understand the importance of being an "Upstander" who pledges to stand up in the face of online shaming. "We assume everyone knows what respect is, but people just don't," says executive director Sean Kosofsky.[8] The chancellor of Rutgers recently read the foundation's Upstander Pledge to all five thousand incoming freshmen, and the pledge has been taken by members of groups from the Peace Corps to Big Brothers Big Sisters.

THE UPSTANDER PLEDGE

I pledge to be an Upstander.

I will stand up to bullying whether I'm at school, at home, at work, in my house of worship, or out with friends, family, colleagues, or teammates.

I will work to make others feel safe and included by treating them with respect and compassion.

I will not use insulting or demeaning language, slurs, gestures, facial expressions, or jokes about anyone's sexuality, size, gender, race, any kind of disability, religion, class, politics, or other differences, in person or while using technology.

If I see or hear behavior that perpetuates prejudice:

I will speak up! I will let others know that bullying, cruelty, and prejudice are abusive and not acceptable.

I will reach out to someone I know who has been the target of abusive actions or words and let this person know that this is not okay with me and ask how I can help.

I will remain vigilant and not be a passive audience or "bystander" to abusive actions or words.

If I learn in person or online that someone is feeling seriously depressed or potentially suicidal:

I will reach out and tell this person, "Your life has value and is important, no matter how you feel at the moment and no matter what others say or think."

I will strongly encourage this person to get professional help.[9]

"I can't think of a behavior that is more universally disap-
proved of than bullying, yet it happens all around us," notes
Kosofsky. "We have a society that rewards aggression; we behave
at a keyboard in ways we never would face-to-face. So much
around us rewards that aggressive behavior. We don't have an
overarching incentive for kindness. There isn't something telling
us, this pays, to be kinder and gentle." Could this be why, despite
years of work being done by prevention programs, online shaming
still seems to be so persistent? According to the Cyberbullying
Research Center, nearly 28 percent of high school and middle
school students over the past decade reported that they have been
victims of cyberbullying—and since 2014, the number of those
reporting being attacked online has risen to one-third.[10] Why has
so little progress been made? "Even though bullying is disliked,
it is seen as a rite of passage," says Kosofsky. "As if it's something
everyone should have to endure because it builds character. It
doesn't build character; it'll destroy your spirit. We have come
to accept that attitude. We have to undo that. What parents and
teachers can do is reverse that thinking."[11]

Kosofsky explains that the foundation's #Day1 campaign
encourages witnesses of online or off-line bullying to do three
things:

1. **Interrupt.** "It is a highly unnatural act—we ask people to
   be courageous. Train your kids to do it."
2. **Report it.** "We have to undo this mentality that you're
   tattling. It's not tattling if you're saving someone's life."

3. **Reach out.** "If you see someone being attacked and
   treated horribly, tell them, 'You should have not been
   treated that way.' It is so vital to get that lifeline."

Sadly, after Ravi was convicted under a New Jersey statute
against antibias crime, he served only twenty days of a thirty-
day sentence, and that conviction was overturned. He ultimately
pleaded guilty in 2016 to attempted invasion of privacy, and was
sentenced to time served. The Clementis responded with a statement calling again for respect: "We call on all young people and
parents to think about their behavior and not be bystanders to
bullying, harassment, or humiliation. Interrupt it, report it, and
reach out to victims to offer support. If this had happened in
Tyler's case, our lives might be very different today."[12]

The Tyler Clementi Foundation continues to work harder
than ever to provide support and effect change. In addition to
offering school and workplace presentations, the foundation
created the Tyler Clementi Institute for Cybersafety, housed at
New York Law School, which offers free legal aid for cyberbullying victims. *Wicked* composer Stephen Schwartz led the creation
of an original choral composition, "Tyler's Suite," which has toured
nationally and performed at the Lincoln Center. The foundation has also proposed the Tyler Clementi Higher Education
Antiharassment Act to Congress, which would require colleges
and universities nationwide to put campus policies in place that
would protect students from harassment or cyberbullying, based
on sexual orientation, gender identity, or religion.

Amanda Todd Legacy Society

"I can't bring my daughter back—but we can keep sharing the message she intended. It's as if she was put on earth for a reason."

—Carol Todd[13]

The gray-toned video diary has been viewed more than 20 million times. She called it "My Story: Struggling, Bullying, Suicide, Self Harm," and the video begins with a silent and faceless girl holding up a series of cue cards, telling in shocking sequence the events of her life story, while the songs "Hear You Me" and "Breathe Me" play in the background.

Only the final card reveals her identity: "My name is Amanda Todd."

The tragic suicide of this fifteen-year-old Canadian girl has touched viewers around the world, propelling her mother, Carol (who we met in chapter 7), to launch a foundation in her honor. "It's not about my kid, it's about what happened to my kid and how we can prevent it from happening to anyone else," Carol told me in an interview.

Many are already familiar with Amanda's heart-wrenching story of the bullying she endured both online and off.

Amanda was an active and typical young girl who did gymnastics and cheerleading, sang in online videos with friends, and loved to write poetry. The turn to her story began in the seventh grade, when she and her friends went online, talking with strangers on chat rooms that were popular at the time. Like happens with many young girls, at one point, a stranger halfway across the world cajoled her into flashing topless into the camera and screen captured the result.

One year later, her nightmare began in earnest. This man contacted her on Facebook and threatened to send around the image if she didn't put on a show for him. Chillingly, the stranger knew everything about her: her address, school, and the names of her family and friends. Amanda tried to put him off, but he made good on his threats. That Christmas break, there was a 4:00 a.m. knock at her family's door from the police, telling them that Amanda's naked photo had been "sent to everyone." In her own words, Amanda says in her video that she was thrown into a stew of "anxiety, major depression, and a panic disorder," followed by a downward spiral into drugs and alcohol.

With the help of her family, Amanda changed schools and made a new group of friends, in an effort to leave behind the bullies. Then the man created a Facebook profile—and used her nude body as the profile picture, targeting students at her new school. As Amanda's video tells it: "Cried every night, lost all my friends and respect people had for me...again... Then nobody liked me, name calling, judged... I can never get that photo back. It's out there forever..."

Amanda changed schools again. Off-line, a bullying incident at her new school only made things worse. Knowing that she was lonely and isolated, a boy she liked led her on and hooked up with her, even though he already had a girlfriend. That girl and her friends cornered Amanda outside school one day in front of a large group of classmates, punching her and leaving her on the ground. Amanda's video tells the rest: "Kids filmed it... I felt like a joke in this world... I wanted to die so bad... When [my dad] brought me home I drank bleach. It killed me inside and I thought I was gonna actually die... After I got home [from the hospital], all I saw on Facebook was...'I hope she's dead.'"

Even after Amanda moved to a different city to live with her mother, changing schools yet again, kids continued to harass her for months, tagging her in photos of bleach to mock her suicide attempt. "Why do I get this?" her words ask plaintively. "I messed up but why follow me?"

On September 7, 2012, Amanda posted the eight-minute video on YouTube, unmistakably another plea for help. "Every day I think, why am I still here?" she asks. The screen cuts to an arm covered with bleeding cuts, and another with the tattooed words "Stay Strong." Her final message, posted in the YouTube description of the video, reads:

> Haters are haters but please don't hate, although I'm sure I'll get them. I hope I can show you guys that everyone has a story, and everyone's future will be bright one day, you just gotta pull through. I'm still here aren't I?[14]

A month later, on October 10, Amanda took her own life.

"I never thought it would happen to my kid," says Carol Todd, speaking four years later. "I've learned a lot from raising my daughter and a lot after her death [about] what we can do better. The two main, key points [are] trust and communication. Teens need to trust [that] they can talk to their parents—or a circle of other trusted adults—without judgment, if something should go wrong for them online. And parents are either not interested, or think they know it all. There's so much education that needs to happen." Carol was never a confident public speaker, but now she is forced to be one. Her schedule is jam-packed, talking to groups about different aspects of Amanda's story, from sexploitation to suicide awareness to cyberbullying, using the imagery of snowflakes to reflect her message—that, like Amanda, we are all unique and yet fragile.

The man allegedly behind Amanda's tormenting was finally captured two years later in the Netherlands, where he was put on trial and convicted for similar charges against thirty-nine other victims, and is expected to be extradited to Canada afterward to finally face culpability for Amanda's case.[15] Although Todd traveled to the Netherlands to see the accused face-to-face, she doesn't want that closure. Instead, what she really wants is to continue the conversation. "Amanda's [video] put her story out there, and every time it's out there, people have a conversation about what happened to her and how we can prevent it. That's what I want—the conversation. I can't bring my daughter back—but we can keep sharing the message she intended. It's as if she was put on earth for a reason."[16]

Dancing Man

> *"Bighearted people far outweigh the small-minded every day of the week."*
>
> —Sean O'Brien[17]

"Spotted this specimen" were the fat-shaming words that sparked the Dancing Man movement.[18]

In March 2015, Sean O'Brien was out with friends at a local bar when he noticed some people laughing and filming him. As a heavy-set fortysomething, he was used to being fat-shamed. But these haters went further than most. They decided to put the video of him up on the website 4chan, writing: "Spotted this specimen trying to dance the other week. He stopped when he saw us laughing." The pair of photos showed a stark before-and-after: in the first, Sean is dancing freely, in the second, hanging his head in shame.

Sean is an unassuming, slightly balding bachelor who works in the accounting department of a hotel outside London. "A friend rang me up and said your face is all over the Internet," he explained in his thick Liverpool brogue.[19] At the time, he had little social media presence other than a LinkedIn account. He found the post and began reading the vile comments. Like many struggling with their weight, he was no stranger to cruel comments, but on the street, he says, "You can ignore it. Once it's said, it's gone," he says. "Once it's online, it's there forever."

The only saving grace was what followed. Appalled by

the cruelty, a group of well-connected LA women, including writer Cassandra Fairbanks and musician Hope Leigh, jumped to his defense, organizing a Twitter campaign they named #FindDancingMan.

"I think the whole world saw that light in his eyes in the first photo of him dancing, and we all collectively saw that light go out in the second photo," recalls Leigh. "Anyone with a heart could feel the shame those people made him feel. I immediately put myself in his shoes and felt motivated to take action. The issue was so clear-cut. Someone made him think he shouldn't be dancing, so the response felt so natural and easy: let's be the someone who makes him feel like he *should* be dancing. That to me just made sense. There also wasn't a ton of time for thought. From that first notion, of 'I'd dance with that guy!,' to the first tweet, to the formal invite, was only a few hours."[20]

Touched, Sean created a Twitter account using the handle @DancingManFound and posted a photo of himself holding a sign that read HI CASS + TWITTER (AS REQUESTED). By May, a GoFundMe campaign had raised enough funds to fly him to LA, create the Dance Free Movement nonprofit organization, and throw a star-studded dance party in his honor at the Avalon nightclub in Hollywood, which was DJed by Moby. The event ultimately spiraled into a four-day extravaganza, where he danced with Meghan Trainor on the *Today* show, threw out the first pitch at a Dodgers game, and was feted by more than a thousand partygoers, including notables from Monica Lewinsky to rock star Andrew W.K. "It was surreal," Sean recalls a year later.

What motivated a group of Hollywood hipsters to reach out to a British accountant halfway around the world? For Leigh, it was that moment in middle school when she herself was teased for developing early that she credits for seeding her with compassion. "I've always been sick to my stomach to see people gang up on others," says Leigh. "As a child, and now online as an adult, it is so easy for the battle to become one thousand to one, and I will always stand up for the one alone. It's overwhelming and unfair, and it has to stop… When we are honest with ourselves, we can recognize where we have been a bully (not outright, but perhaps through exclusion) and where we have been bullied ourselves. It isn't ending with youth. We still face it as adults."

Perhaps that's why this story also struck a chord with so many of us.

"Sean's party wasn't just about him," Leigh says. "He worked hard to let the party be a celebration of everyone [who] had been marginalized, as a beacon, or reminder of hope, positivity, and acceptance, for everyone to know that there are people out there who will want you to dance… Though we never expected it to receive that amount of attention and support, I am happy to have been a part of it for the joy it did bring."

Sean believes it's because his story turned from a tale of sorrow into one of inspiration. "There's good in the world," he says, "and it far outweighs the bad."

Do you have a hero story that's making a difference in your life or community? Maybe it's time you share it with others. In the next chapter, we will find out how we can all be part of making our digital space a kinder place.

CHAPTER 10

TAKING ACTION

"Carry out a random act of kindness with no expectation of reward, safe in the knowledge that one day someone might do the same for you."

—Princess Diana[1]

What can we, as parents, coaches, educators, or simply concerned citizens, do to bring civility back to public life? How can we all share more mindfully—but also take care not to shame others for the mistakes they have made or for just being who they are?

Create Empathy

Perhaps the very first place to start is with a renewed emphasis on teaching empathy to our children. Parenting expert Michele Borba, EdD, author of *UnSelfie: Why Empathetic Kids Succeed in Our All-About-Me World*, makes a strong case that a decrease in basic empathy has created a culture ripe for online attacks. Researchers at the University of Michigan crunched data that tracked years of incoming college freshmen's empathy, and found that empathy has declined by 40 percent in the last three decades—while narcissism has risen by 58 percent.[2] This inability to see those on the other side of the computer screen as people deserving of our compassion is one of the huge drivers of our Shame Nation. "Depersonalization is what's happening," says Dr. Borba. "In a lot of these [cybershaming] cases, the person is hundreds of miles away; you'll never be face-to-face. It becomes an easy click."[3]

Can we reverse course and make sure that we are passing along empathy to the next generation? Yes, Dr. Borba emphatically believes. "This is a human disaster, not a natural disaster," she says. "We caused it, we can turn it around. Empathy can be cultivated; we just have to work it into our parenting agenda." (And that doesn't just apply to our children. There are empathy-boosting courses available for adults too. About 20 percent of employers in the United States, such as LinkedIn, Cisco Systems, and Ford Motor Company, have offered empathy training for managers, according to the *Wall Street Journal*.[4])

Modeling works as well. "Your teen cares about what you think. They will also watch your behavior online," Dr. Borba

says. "It's imperative that parents today are conscious of their own posts and comments. It's not just about them—they have an audience that looks up to them."

In *UnSelfie*, Dr. Borba lays out nine specific ways that we can revitalize empathy in the upcoming generation and in ourselves. We can work to improve the ability to recognize emotions in others, instill a moral code to become a caring person, and learn to take on another person's perspective.

With greater empathy and compassion, it should be impossible to leave cruel comments. To get started, let's adapt Dr. Borba's four-step method, which she calls CARE, toward how we approach posting online.[5]

C = Call Attention to Uncaring. Did you notice that there was an ugly comment on someone's post? Was it about you? Talk about it.

A = Assess How Uncaring Affects Others. Was your teen a victim of a cruel comment, or were you? Discuss how this made you feel.

R = Repair the Hurt and Require Reparation. Did you or your teen write a comment that hurt someone (even if you didn't mean to)? Immediately delete that comment, apologize, and contact the person personally.

E = Express Disappointment and Stress Caring Expectations. We're all human, and we're going to make mistakes. It's what we learn from them that matters. Be a caring and kind role model at all ages.

Galit Breen, the writer fat-shamed for her wedding photo, turned around and penned the book *Kindness Wins*, a call to arms for greater kindness online. In her 2015 TEDx Talk, "Raising a Digital Kid without Having Been One," she pleads with parents to emphasize this trait when raising their kids. "We hold in our hands the missing piece between the good, kind, smart kids who we know and love, and those very same kids who are being so reckless with themselves and with each other online. That missing piece is short, direct, repeated, ongoing conversations, not about how to become bully proof, but about how to make sure that they're the ones who are not doing the bullying... We can teach them that there is a difference between intent and impact. How loud and permanent the Internet is, and that there is no such thing as online privacy, but there is a difference between fighting issues and people. And we can teach them that on the other side of every single interaction that they have online is a real human being."[6]

Trust, But Confirm

While we want to trust our children, parents also need to make sure that they are checking in and checking up. Tools such as monitoring software and antibullying apps do have their place, but there is no substitute for ongoing dialogue with children about their online lives. I often remind parents that a child's cyberlife is constantly evolving. Parents have to learn to ask their children on a regular basis about their digital lives. Are there any new apps they've downloaded? Websites they've visited?

Cyberfriends? It must be as common a question as "How was your day at school?" One study revealed that in only one out of four times a teen encountered a risky online situation did they tell their parents about it.[7] It's important to keep your lines of communication open so if your child is being harassed online, he or she feels comfortable to come to you. Being interested in their online activity is as important as being interested in their education—it's their life today.

Social-media-savvy police officer Mike Bires says this: "It should be made very clear, at the beginning of their 'digital experience,' that Mom and/or Dad will always be checking on their activities until they turn eighteen. There are no exceptions, no debates, and no way around this. *Period.* 'Isn't this an invasion of their privacy?,' some may ask. My response to those who feel their children are entitled to privacy is that those parents need to deal with whatever comes [from] that belief. Don't come to law enforcement wanting us to fix the problem you had the ability to prevent."[8]

Officer Bires reminds parents that if their child becomes a victim of sexting or cyberbullying, as a police officer, this is his best advice:[9]

- Be empathetic, honest, and transparent.
- Discuss the incident with school administrators (if applicable).
- Preserve the evidence: do not delete any communication with the offender.

- ✦ Report the incident to law enforcement.
- ✦ Seek professional counseling for both you and your child to manage the incident.

If you are dealing with online harassment and are a victim of cyberbullying, Bires offers some additional advice for youth:

- ✦ **Do not engage.** Ignore the bullying, however difficult it may be. When you do so, the bully's desired reaction will not occur, and the bully will end up looking worse among his peers. The bully will quickly find that those he thought would support his actions will turn away from him.
- ✦ Speak positively online and never show your frustration or anger. It's easier said than done, but do your best.
- ✦ Block, restrict, or limit the offender's access to your conversations. For example, on Facebook, you have the ability to block users in your settings area. Many platforms allow you to block people to eliminate them from engaging with you on social media.
- ✦ Reach out to your **parents** or a **trusted, responsible adult.** Seek their assistance in dealing with the issues. There is absolutely nothing wrong with getting help from others who might have gained some experience through similar circumstances.
- ✦ Report the offenders to the social media platform where they are committing their acts. Those platforms have conduct rules established for users of their platform, and

they will take action if necessary to alleviate a person's frustration with using their platform.

+ If it persists, seek the assistance of law enforcement.

Community Kindness

Does your school, neighborhood, or community have a problem with bullying? Can we all find a way to instill kindness that will hopefully carry over into our online lives? Many local and international organizations have sprung up with this exact mission, while motivated individuals are also finding ways to insert a note of compassion into the local dialogue.

The Ripple Kindness Project

Founder Lisa Currie developed this Australian-based curriculum for the very youngest of students in elementary schools. "From what I could see, traditional antibullying programs were very negative and short-lived and really didn't leave kids with any resources for improving their thoughts, feelings, or behavior," she says. "Our aim is to infuse children with goodness by teaching them about their emotions, having them participate in acts of kindness, and experiencing the good feelings that are produced when they do good for others. When children learn to be givers, their whole world can change."[10]

Here are some of Ripple Kindness's suggested activities to get you started:

- Give blood.
- Leave a chocolate for the cashier.
- Bake a cake for someone.
- Feed an expired parking meter.
- Pay for someone's meal.
- Give a compliment.
- Listen to and play with children.
- Clean someone's home.
- Clean someone's car.
- Buy coffee for the person behind you.
- Visit someone in a nursing home.
- Take some food or clothing to a homeless person.
- Leave a note in a lunch box.
- Don't charge someone for some work you do for them.
- Become an organ donor.
- Ask an elderly neighbor if she needs any assistance around her home.
- Hand make cheer-up cards and deliver them to a hospital for patients.
- Let someone go in front of you at a checkout.
- Babysit for someone.
- If you're an employer, allow your staff to leave an hour early one day.
- P.S. It's always fun to pay a toll for someone behind you. It's about paying it forward. Let the attendant know that your only request is for the person to pay it forward someday.

I Can Help Delete Negativity

At Excelsior Middle School in Byron, California, a fake Facebook page was created to poke fun at a teacher. Leadership teachers Matt Soeth and Kim Karr were inspired to create a program called #ICANHELP, empowering their middle school students to choose a different path and "delete negativity." They call themselves the Positive Warriors. When, a year later, a similar page went up on Instagram mocking the same teacher, thirty kids went in, posted supportive comments, and reported the page, which was removed within forty-five minutes.

This crusade is spreading. #ICANHELP visited more than one hundred schools across the country, hosting assemblies and organizing student leadership trainings. "I don't tell kids what not to do, I show them what I want them to do," Soeth says. "Modeling that expectation will breed that behavior."[11] One popular exercise that Soeth brings to schools is "Give a Compliment, Get a Compliment," where students scrawl personal messages for friends on Post-it notes, filling up entire bathroom mirrors with notes like "You're amazing," "Smile!," and "You're the best version of you!" Another is "High-Five Highway," where high school upperclassmen and staff line the halls on the first day of school, and instead of hazing new freshmen, greet them with high fives. The most popular of all is "UnSent"—if you've ever wanted to thank someone for something they did, but fretted that too much time had passed, UnSent Day gives you the chance to still send that missive long past its due.

"If there's negativity online, there's going to be negativity within the culture off the campus," says Soeth. "We try to work with students to [help them] understand how powerful they are, in order to make good things happen."

The Kind Campaign

After experiencing bullying as adolescents, film school graduates Lauren Paul and Molly Thompson set out to produce a documentary exploring mean-girl behavior. Their award-winning 2009 film, *Finding Kind*, would eventually evolve into something of a movement known as the Kind Campaign. To date, the group has held more than a thousand free school assemblies, beginning with a scene straight out of the film *Mean Girls*, where girls are asked to raise their hands if they've ever been hurt by something another girl has done to them—and then leave their hands in the air if they've ever done something to hurt another girl. "It's pretty amazing to see," says Molly. "They realize they're all on same page. They have this aha moment."[12]

After watching *Finding Kind*, the girls are inspired to get up and publicly apologize for a misdeed, pledge to make a change, and write a note of appreciation to an acquaintance. Girls are driven to tearful confessions in the "Truth Booth" about how emotionally scarring bullying can be. "We try and leave girls with the message that [the bullying] they're experiencing is one small, tiny chapter of their whole story," says Molly. "It's so hard to really grasp that there are so many amazing experiences they will have, [that] they don't need to change themselves."

One story lingers with both founders: a girl named Rachel

approached them after one assembly and confessed that she had
woken up that morning contemplating suicide. After the program,
she posted these words on their Instagram account:

> I came to school today on the verge of tears. I came up and
> shared my Kind Card. Once I got home I realized that even
> though some girls can be mean and don't understand what
> their words can do, that it shouldn't be worth dying over. Your
> assembly today? Yeah, it saved my life.[13]

"It was a beautiful and encouraging moment for us," Lauren
smiles. "Now she's graduated, doing well, and we still keep in
touch with her." More than 450 schools have already started up
their own Kind Clubs, which are like Girl Scout troops but on a
mission for harmony, with tasks such as performing one random
act of kindness per week. Might your school be next?

#NiceItForward

"Who is the most awesome person today?" asks the Facebook
page "Greenwich Compliments."[14] And every day, it answers,
peppering those who live in this posh Connecticut town with a
daily hip-hip-hooray for their beautiful voice, sense of style, or
willingness to lend a hand.

The idea was sparked after a 2013 suicide of a bullied teen on
his first day of his sophomore year shook the entire community.
Looking for a way to turn things around, the Facebook page solic-
its compliments about Greenwich residents, receiving up to thirty

submissions daily. "It only takes a few seconds, but by submitting, you are making a conscious decision to make someone else's day better," says the woman behind the site, a Greenwich High School graduate who chooses to remain anonymous. "People ask if I am a police officer or a teacher. I am neither. I am just a person who grew up in Greenwich and who knows how tough life can be sometimes and who knows how awesome it is to receive a compliment and how rewarding it is to give one."[15]

"Privacy and anonymity are very frequently used negatively on the Internet, unfortunately," she explains. "I like to think that Greenwich Compliments is different."

These types of #NiceItForward accounts have popped up across the nation, some created by students themselves and others by adults. One Twitter account, @OsseoNiceThings, was created in 2012 by Kevin Curwick, then a popular high school senior at Minnesota's Osseo High School, who began anonymously tweeting shout-outs to his classmates. The media attention sparked similar accounts all over the state and beyond, such as @ERHSnicewords, created by a student at East Ridge High School in Woodbury, Minnesota, and titled "The End to Bullying," which now has more than forty thousand followers.

Down in Charlotte, North Carolina, one father of two had had enough with comments on Twitter bashing a sports media star he follows, so he decided to create the moniker Supportive Guy and put positivity out there. "I wondered what the counterpoint of this kind of online behavior would be," he explains. "And I created the account [@SupportiveDude] that moment on a whim." Since then, he has grown his following to more than fifteen hundred fans, tweeting kind remarks, and has even started the *Supportive Guy* minute-long podcast. "All I'm trying to do is give people a safe haven and an online friend they can always tweet and get a response [from], maybe even a laugh. And maybe, people can see that you can get attention online without being negative. The reality is that you have a right to be on social media whenever you want to, without [the] risk of being verbally attacked."[16]

Greenwich Compliments echoes this notion. "I have this view that we fail to realize how much our words matter, how impactful they can be, both positively and negatively," says its founder. "I am really interested in this idea that every comment we say about someone else is a seed. We plant that seed, that idea in that person's mind, and then it grows, fueled or watered by other comments or experiences that person has. Negative comments, negative thoughts, they tend to grow like weeds. For example, I say you aren't athletic, and the seed is planted. Then in gym, maybe you are the last guy picked for a team—that seed starts to grow. Then you strike out [in baseball] and someone makes a comment. That single idea just keeps growing. Positive comments are similar; they just grow a lot slower. For some reason, it is much easier for you to believe me when I say you are ugly than it is when I say you are beautiful. So I guess that's the idea—to combat those negative thoughts we all have about ourselves by planting positive ideas in people's heads. We need to start focusing on everyone's triumphs instead of their flaws. We all achieve great things but rarely get the proper recognition; we as a society are just too focused on the bad. I don't necessarily believe that one compliment is enough to stop that weed from growing completely or from removing it altogether, but if it helps even a little bit, then it is worth it."

Could your community use a #NiceItForward campaign of its own?

The Toothpaste Challenge

It's debatable who first came up with the idea of the Toothpaste Challenge, but it's the perfect way to send a strong message that what we say can't always be taken back. One mother, Amy Beth Gardner, posted on Facebook her own use of the powerful demonstration with her own daughter, Breonna, the night before she started middle school.

"I gave her a tube of toothpaste and asked her to squirt it out onto a plate. When she finished, I calmly asked her to put all the toothpaste back in the tube. She began exclaiming things like, 'But I can't!' and, 'It won't be like it was before!' I quietly waited for her to finish and then said the following:

> You will remember this plate of toothpaste for the rest of your life. Your words have the power of life or death. As you go into middle school, you are about to see just how much weight your words carry. You are going to have the opportunity to use your words to hurt, demean, slander, and wound others. You are also going to have the opportunity to use your words to heal, encourage, inspire, and love others. You will occasionally make the wrong choice; I can think of three times this week

I have used my own words carelessly and caused harm. Just like this toothpaste, once the words leave your mouth, you can't take them back. Use your words carefully, Breonna. When others are misusing their words, guard your words. Make the choice every morning that life-giving words will come out of your mouth. Decide tonight that you are going to be a life-giver in middle school. Be known for your gentleness and compassion. Use your life to give life to a world that so desperately needs it. You will never, ever regret choosing kindness.[17]

Amy's post was so powerful, it was shared nearly a million times. What if we all took the time to make our kids take the Toothpaste Challenge?

Positive Slamming

Sometimes what begins as a shaming can flip, as supporters show you they have your back. This is called positive slamming, and we can all be more proactive when we see a cybershaming underway. Tennessee high school senior Kristen Layne decided to try to sell her junior prom dress, a sparkling purple strapless, in order to buy a brand-new one, with the help of a GoFundMe page.[18] In spring 2015, she posted a picture of herself in the gown on a local Facebook yard sale page, "For Sale in Sumner County." To her surprise, her post began garnering fat-shaming insults on Facebook from a handful of cruel male strangers in the group.

"Can you please stop with the comments?" Kristen responded. "Sorry that I'm not pleasing to your eye." Within minutes, she found that hundreds of strangers had jumped to her defense. "It's obvious this woman is beautiful, inside and out," wrote one poster. Within a few months, instead of raising $350 for a new dress, Kristen's page had received $5,000 in donations of support—and she bought her best friend a prom dress as well.[19]

Have you seen online criticism in your own community? Are you ready to step in with words of encouragement when necessary? It's hard for victims to stand up for themselves, but when others reply, it tells the bully that people are watching and that the victim has supporters who won't tolerate abuse.

You can model your response after one group of parents in North Pocono, Pennsylvania. They were alerted by high school officials that there was intense bullying going around on the latest anonymous app being used by their teens, Burnbook, the brainchild of Jonathan Lucas, developer and CEO. These moms and dads organized a campaign to step into the conversation with cheesy but positive messages of acceptance each morning, like "Be kind to each other!" and "Remember to be positive. NP Moms & Dads."[20]

Teaching our kids to speak out when they see online bullying is a worthy goal, but inserting themselves into the conversation can be tough for many kids—and even adults—to do, according to Sameer Hinduja, cofounder of the Cyberbullying Research Center. "In the real world, bystanders are going to be hesitant, because they don't want to be targeted," he says. "A lot just want to mind their own business and not get involved." Hinduja has an alternate suggestion: "Just check in on the people who have been targeted, drop them an encouraging text, just love on them and care about them, and repeatedly check on them. I think that goes a long way. When I was being targeted [by bullies in school], I just wanted someone to say, 'I'm still in your corner.' That's all that I really needed, and I knew I could survive."[21]

If you're nervous about being an Upstander (to use the Tyler Clementi Foundation's term) on your own, you could also try getting others involved. That's what happened at a high school in Hillsborough, California. Freshman soccer goalie Daniel Cui was blamed for failing to make a game-saving block and for his team's abysmal record. Photos of him were posted, shaming him as the WORST GOALIE EVER. In response, a few friends found a photo of Cui making a save, and together they each posted it as their Facebook profile picture. Then the entire boys' soccer team joined in, then the girls' team, and ultimately, a majority of his classmates joined in. This simple act of solidarity restored Cui's confidence, and, as legend has it, he came back to help the team to a win the following season. "The whole school had stood up for someone who needed it," recounts the narrator of an uplifting Facebook Stories video. "He was a normal kid just like us. We have our highs and our lows, and that's when we realized…that we were all Daniel Cui."[22]

Stay Skeptical

When you do see something nasty posted about someone else, don't always assume it to be true. According to research published in the journal *Computers in Human Behaviour*, nearly two out of three people admitted to lying online, even on social media sites like Facebook where they use real names, and more often on dating sites like Tinder or anonymous chat rooms.[23] Is that any surprise? Of course we tell small white lies about ourselves, to make ourselves look better online, especially on a dating app.

But did you ever stop to think whether it's possible that something *negative* posted about someone could be false as well? Be aware that those with personal vendettas or business interests could be deliberately sabotaging someone's reputation.

"This is happening more and more," says attorney Christina Gagnier, who tells of one client who met a woman on Tinder and went on four dates before the relationship petered out. This angry woman set out to destroy him online, posting photos on ReportMyEx.com, creating defamatory Tumblr and WordPress pages, and even distributing pamphlets to his professors and on neighbors' cars. Gagnier took on the case—and in return, had her own reputation trashed on a complaint site.[24]

In Boston, one local café found itself receiving a stream of targeted one-star restaurant reviews after the sous chef posted a video documenting disturbing talk and imagery at a local pro-Trump rally on his personal Facebook page.[25] And remember the author from chapter 7 whose book received one-star ratings on Goodreads after she expressed her political views?

"You hope that people see through the maliciousness of [the negative reviews]," worried the café owner in a *Boston Magazine* interview. But do they? A survey by YouGov found that even though 90 percent of Americans said they didn't fully trust online reviews, they still relied on them to make purchasing decisions.[26] Shockingly, one in five people surveyed admitted to writing false reviews of products they'd never actually tried. Remember, small business owners have almost no recourse to these slams, since opinion is a protected form of speech. In July 2016, a Yelp VP announced on the company's blog that if any small business threatened to fight a negative review with legal action, Yelp would place a pop-up ad on its page accusing the business of stifling free speech.[27]

"We really have to be careful with doing Internet searches because not everything online is true," says attorney Bradley Shear. "There's a lot of false information, bad online reviews where companies have paid to have bad things said about competitors. I think more people are beginning to say, 'I've got to take everything I see online with a grain of salt.'"[28]

How can we sort through what we read online, especially given what we now know was a spike in fake news articles widely circulated during the 2016 presidential election season? Internet critic Howard Rheingold, author of *Net Smart: How to Thrive Online*, created a quick way to evaluate anything you see on the web for accuracy, using four criteria. It's known as the CRAP Detection test:

- **Currency:** How recent or up-to-date is the information?
- **Reliability:** Is the content opinion-based or balanced? Does it provide references and sources for data?

+ **Authority:** Who is the author or source, and are they reputable?

+ **Point of View:** Does the poster have an agenda or is he or she trying to sell something?[29]

Finally, be sure that what you are passing along is equally credible, advises tech journalist Larry Magid. "What comes out of our fingers is the same as what comes out of our mouths. Be responsible. If you are sharing a post that is wrong or mean, you're saying that. If you're liking it, you're endorsing it. I think people need to think about what they say and what they pass on. Is this something I know is true? Would I be proud, if I said these things out loud?"[30]

Educate in Schools

Thanks to antibullying legislation now passed in all fifty states, many school administrators have a mandate to bring some form of programming to their schools. "Some [antibullying laws] are weak in our opinion and some are solid," says the Cyberbullying Research Center's Sameer Hinduja. For example, he says, Montana simply requires schools to have such a stated policy in effect, while states like New Jersey and New Hampshire go further, specifying prevention components that must be implemented, such as educator training, student assemblies, and an anonymous reporting system. "I would love all states to mandate that as law." (To check your state's law, visit cyberbullying.org/bullying-laws.)

Even with mandated programs, many schools struggle with how best to incorporate these lessons, which range from cyberbullying to sexting. Do they belong in health class or computer skills? Some schools rely on computer programs, like Common Sense Media's Digital Compass, which takes students through Choose Your Own Adventure–style animated scenarios of positive and negative Internet safety choices. Others schedule one assembly a year, bring in a cyberexpert with a few horror stories, and call it a day. "Kids completely tune that out," says #ICANHELP's Matt Soeth.[31] He is backed up by the findings in Cybersmile's 2015 Stop Cyberbullying Day annual report, in which the number-one thing (at 33 percent) that teens said would make them kinder to others online was reading real-life stories about those who have been affected by cyberbullying. The least effective, at 3 percent? A lecture from a police officer.[32]

Another alternative is the innovative program created by Cyber Civics, a California-based start-up that has developed a three-year comprehensive curriculum to teach all aspects of digital literacy to middle schoolers. After a cyberbullying episode hit her daughter's public charter school, parent and video producer Diana Graber developed this program based on the master's degree she had just received in media psychology and social change. Graber still teaches the course herself, but also trains teachers to run the program at their own schools, providing video and written materials for a fee. Since its inception, the program has grown to be offered in more than one hundred schools in twenty-five states and overseas, ranging from Waldorf to public schools.

Sixth graders begin with the basic concepts of digital citizen-ship, covering digital footprints, what should never be shared online, and antibullying behavior, such as the difference between being an upstander and a bystander. Seventh graders focus on research skills, covering concepts such as keywords, Wikipedia, fair use, browsers, search engines, and privacy protection. By eighth grade, the students shift focus again to consuming versus producing online content, covering media literacy issues from sexting to Photoshopping to copyright protection. The final exam is a series of questions we adults would likely fail: What are cookies and how do they work? What does *URL* stand for? What is a spider? What are the eight tips for a secure password?[33]

While much of the same information is on her comple-mentary website, CyberWise.org, Graber ultimately found that approaching the students directly, instead of using their parents as mediators, works best. "Kids don't want to talk to their parents in middle school," she says. "The talking is with each other. If we can make safe spaces in the classroom, that is way more powerful." Graber knew her message was received when a new girl posted a photo of herself in a bikini, and an eighth-grade boy who'd taken the course scolded her. "You need to take that off your Instagram," he told her bluntly. "That was stupid." Harsh, Graber concedes— but effective. "In a crude way, he was looking out for her. The kids start being each other's mentors."[34]

If a cyberbullying incident erupts on campus, one new resource that school administrators nationwide can to turn to is the iCanHelpline.org, a social media helpline run by the nonprofit

The Net Safety Collaborative, a merger of #ICANHELP and Net Family News, Inc. If a student has received no response within forty-eight hours of reporting abuse to the platform, hotline helpers can escalate the situation using their tech company contacts. "There's a huge lack of understanding on the education side [of] how, and even that you *could* get content removed," says Soeth.[35] One principal told him about a page that was up for two months because he was unaware that reporting was an option. Launched during the 2016–17 school year, after piloting the program in California, Washington, and Georgia, the helpline is the first of its kind in the United States.

Advocate

Consider becoming an advocate for this cause: talk to politicians about creating or refining laws to address twenty-first-century online harassment. In 2014, the European Union passed the Right to Be Forgotten law, which permits people to petition to have specific web results about them removed from a search engine. Google reported that between May 2014, when it implemented its official request process, and February 2017, it received nearly seven hundred thousand requests, flooding in from Europeans eager to have information about themselves on sites like Twitter, Facebook, and YouTube scrubbed from search results.[36] In 2016, as part of a new code of conduct designed to combat hate speech, Facebook, Twitter, Microsoft, and YouTube agreed to European regulations that require them to review "the majority of" hateful online content within twenty-four hours of being notified—and to remove it, as needed.[37] Should America consider similar laws? I personally strongly believe in free speech, but that doesn't mean it's okay to defame, harass, or threaten another person.

Our fundamental First Amendment principles present a uniquely American problem for crusading against online harassment. "In theory, the Right to be Forgotten makes a lot of sense," says security expert Theresa Payton. "However, one person's 'right to be forgotten' is another person's 'right to be remembered.' What if you and another person are quoted in an article, and you love that quote, but the other person wants it taken down because they received negative feedback? Who decides? Do they merely strike the one quote and leave the other? I think a lot of lessons can be learned, and I'd like to see social media and search engines make it easier to request removal of items."[38]

Larry Magid agrees that the Right to Be Forgotten is a very difficult concept to reconcile with our free-speech culture. "What Europe doesn't have is a First Amendment. I fully understand how people who made a mistake want to, and maybe should, have that expunged from their record," he says. "On the other hand, people are responsible for their actions. People having to relive images [like the family of Nikki Catsouras], it's horrible. But it's part of the price of living in a society like ours. I tend to err on the right to free speech."[39]

In her book *Hate Crimes in Cyberspace*, law professor Danielle Keats Citron lays out several potential changes to the U.S. legal system that could strengthen the recourse for victims of revenge porn or online harassment, such as the passage of a federal law that would make posting nonconsensual pornography a felony, removing the current legal immunity (under the Communications Decency Act) for operators of sites that are dedicated to revenge porn, and allowing victims to sue using a pseudonym to protect their privacy.[40]

Unfortunately, the wheels of government grind along slowly, and while these proposals languish, people's lives are being damaged. "It takes years for the law to catch up with technology," admits attorney Bradley Shear. "Those looking to quash anonymous critics or stop doxxing forget how strong the First Amendment is. There hasn't been a satisfactory way for the law to deal with these issues."[41]

Why not take matters into your own hands and run for office yourself? Brianna Wu was so infuriated by her Gamergate online harassment experience that she announced a 2018 bid for a seat in Congress.

Nonconsensual Porn, Sexting, and Sextortion

At press time, thirty-four states now have laws against nonconsensual pornography—if you're in one of the remaining states that doesn't, consider adding your voice to that cause. Or join the push for federal legislation to make revenge porn a crime nationwide. This bill, known as the Intimate Privacy Protection Act, was introduced in July 2016 by Congresswoman Jackie Speier of California and would penalize anyone knowingly distributing a sexual or nude image of someone without that person's consent, with a penalty of up to five years of prison time. Importantly, because this would be federal law, similar to copyright protection, technology platforms like Google would not be shielded by the legal immunity they enjoy under the Communications Decency Act. Instead, they would need to remove the material in a timely manner or face liability, while sites that intentionally promote or solicit the material will be subject to prosecution. Speier vows to continue to press the bipartisan legislation until it passes. "With the onset of the Internet and social media being so ubiquitous," she says, "the potential for personal lives to be so severely impacted by the distribution of nonconsensual pornography really required legislative action."[42] Free-speech advocates such as the American Civil Liberties Union (ACLU) fret over the potential implications, but the law was crafted to include certain exceptions, such as a compelling public interest or if the photo was taken in a public space. The United Kingdom passed its own law making revenge porn a crime in 2015 and has already prosecuted more than two hundred cases in its first year of enactment.[43] When will we?

Sexting laws for teens could also use some rehabbing— should two teens swapping photos with each other really be at risk of being charged with a felony and placed on a sex offenders list for life? Does your state need to refine its sexting laws as they apply to minors? (Check your state's current laws at cyberbullying .org/sexting-laws.) California, for example, is considering a new statute that would allow schools to expel minors who sext, but not label them predators for life. Other states have chosen to make this infraction a misdemeanor with punishments of community service.

Laws are also needed to tighten federal regulations against sextortion, where there is currently a gap that leaves adult victims vulnerable. Finally, nursing homes and hospitals need to implement stricter prohibitions to prevent their staff from disseminating photographs of their patients without their consent. Has it become necessary to require medical personnel and paramedics to sign confidentiality agreement paperwork that protects patients' privacy?

Cyber Crimes

U.S. congresswoman Katherine Clark of Massachusetts has positioned herself as one of the leading advocates of the fight against cyberharassment. Clark, a former lawyer and the parent of three teenage sons, was drawn into the battle after pleas for help from Gamergate victim Brianna Wu, who lives in Clark's district outside Boston. Wu and her husband were forced to flee their home following online threats to her life. "Graphic threats of rape, murder, and harm to her and her husband included the specific times of day these crimes would occur, what weapons would be used, and even released her private home address," Clark says. "On a separate occasion, a bomb threat directed at her business forced her to withdraw from a popular video game conference."[44]

While the Department of Justice has laws that can be used to prosecute these crimes, Clark concluded that, sadly, many times they are not enforced. One of her first proposed bills gave the FBI a simple mandate to make pursuing these cases a higher priority. "The work we've done since our meeting with the FBI has been aimed at ensuring the Department of Justice has the resources and guidelines necessary to enforce the laws that are already on the books banning the worst types of online abuse, as well as making sure we're closing gaps in laws that haven't caught up to abuses like sextortion and swatting."

As discussed in chapter 8, too many victims reporting their crimes find law officers to be clueless, dismissive, or both. Famously, female journalist Amanda Hess wrote of her experience being on the receiving end of threats of rape, and after reporting them to the police, getting asked by an officer, "What is Twitter?" In response to this sort of ignorance, Clark has introduced the Cybercrime Enforcement Training Assistance Act, recognizing that many police officers and departments lack the technical skills and manpower to track down cyberfoes. The bill would allot $20 million in annual resources and training and also establish a National Resource Center on Cybercrimes Against Individuals to track such cases.

"I think that police officers want to help," Clark told *Elle*, but "they just don't have training. We need everybody to be able to know how to do a basic forensic investigation online."[45] In September 2016, Clark proposed an additional bill: the Cybercrime Statistics Act, hoping to force the federal government to do a better job of tracking instances of online crimes against individuals, such as stalking, harassment, or other attacks. (Police officer Mike Bires goes further: he'd like to see an entirely new federal agency dedicated to crimes involving the Internet and technology.)

As with most advocates against harassment, Clark's

outspokenness has put her squarely in the line of fire. After she introduced the Internet Swatting Hoax Act, making these potentially lethal swatting pranks a federal offense, she and her family found themselves on the receiving end of such an attack one Sunday night at her Melrose, Massachusetts, home. "No one should ever have to answer her door to the police in the middle of the night just because someone disagrees with her," Clark says. "These hoaxes are not only dangerous to innocent civilians, they're also dangerous for our law enforcement and costly to local police departments." As she told *Mother Jones*, "The intent was probably to make me back off, but it certainly made me more committed to continue to try to get legislation passed."[46]

"The biggest thing people can do to help us pass our bills is to make sure they let their representatives know how important ending online abuse is to them," Clark added when interviewed. "When people take a stand against online abuse, share their own stories to let others know they're not alone, or push their lawmakers to address these practices, we make progress."[47]

Can Tech Save Us?

> "We suck at dealing with abuse and trolls on the platform."
> —Dick Costolo, former CEO of Twitter[48]

Technology got us into this mess. Can it also get us out?

Innovative methods of fighting cybershaming with high-tech solutions are being developed daily. Hack Harassment, a newly formed initiative founded by Intel, Vox Media, and Lady Gaga's Born This Way Foundation, is attempting to fight online harassment by partnering with tech companies to find technological solutions.

Civil, a start-up based in Portland, Oregon, aims to restore "civil discourse" to news media message boards. Its Civil Comments product uses a peer-review system that analyzes users' votes on each other's comments and pulls back abusive comments.[49] Google's in-house incubator, Jigsaw, has developed the software tool Conversation AI, an "automated harassment detector," which the *New York Times* plans to use to help flag and block potentially abusive online comments until they can be reviewed by human moderators, according to a piece in *Wired*.[50]

Some advocates believe that data can solve the problem of online harassment. Mary Aiken, the Irish forensic cyberpsychologist who inspired the CBS television series *CSI: Cyber*, writes in her book, *The Cyber Effect*, of her ambitious plans to create the Aiken Algorithm. This code could use big data to flag active cases of online harassment and send a digital alert to parents, or a supportive message to a young person being bullied. She calls this the "math of cyberbullying": who is sending the messages to whom, multiplied by the hateful words and the escalation and frequency of attacks.[51] Aiken has put out a call for anyone who has ever received a hateful message to forward it to her repository, where she will crunch the numbers to help create this novel use of artificial intelligence.*

---

* Visit www.cypsy.com to contribute to or learn more about this project.

App developers are also jumping into the battle. Here's a sampling of recent innovations:

- **Amanda**, an antibullying app developed by a team of Greek designers in honor of Amanda Todd, took first place in the World Citizenship category of Microsoft's prestigious Imagine Cup competition.[52] Amanda is designed to capture those with bullying "tendencies" and intervene.

- **ReThink**, a software program created by teen entrepreneur Trisha Prabhu and featured on *Shark Tank*, flags harassing messages and attempts to get a teen to "rethink" it before actually posting it on social media.[53]

- **Sit With Us**, an app invented by sixteen-year-old high school junior Natalie Hampton, from Sherman Oaks, California, takes a modest but inspired approach: it helps high school students easily find a welcome table to join in the otherwise intimidating hierarchy of the high school cafeteria.[54]

- **Block Together**, an app that allows Twitter users under attack to share lists of blocked users with one another, or to automatically block suspicious accounts, such as those that are newly created or have very few followers.[55]

While these innovative start-ups are a positive first step, we can also lobby established Internet and tech companies to strengthen their resolve to fight against cyberharassment. Legal scholar Danielle Keats Citron suggests some intriguing and innovative reforms that tech companies could make, such as unmasking the "true identity" of someone who has made offensive posts, forcing them to stand behind their identity, and offering victims reduced pricing on ad space to rebut their attackers.

Despite making some changes, Silicon Valley has been slow to champion this cause. "The first step is, platforms could care," says Lindsay Blackwell, who worked in the tech industry before leaping to academia, in respect to the issues surrounding online abuse and harassment. "It's easy to see why they don't prioritize the experiences of one user. If 90 percent are having a swell time, and a marginal percentage are experiencing these horrible things, it's difficult to devote financial resources to that problem. As a human being, I don't get it."[56] As Nancy Jo Sales concludes in *American Girls*, "While individual social networking sites have responded with increasing monitoring of users' posts and comments, often when under fire for some story in the news involving abusive behaviors online, there's been no general reckoning in Silicon Valley about the ways in which its products may be encouraging unethical and harmful behavior. The First Amendment has become the blanket behind which social media companies seem to hide from any questions regarding the online speech and activities of their users."[57]

One place many cyberbullying experts say to start is keeping young kids off social media. CyberWise's Diana Graber points out how ludicrous it is that the tech industry can data mine us to death, but can't figure out if a nine-year-old is using Facebook. "Some of our troubles would go away if we all respected age limits on social media," she says. "Developmentally, by age twelve, [children] have the capacity for abstract thinking [and] thinking through consequences, but before that, they can't. The industry should do a lot more."[58]

Adults need some protection too. "Social media platforms specifically need to take enforcement of social norms seriously," says Ari Ezra Waldman, director of New York Law School's Innovation Center for Law and Technology. "A platform rife with harassment of minorities is a platform that silences minorities, and nothing good comes from that. They need to decide what kind of environment they want and to use both [artificial intelligence] and human tools to ensure, as best they can, that such an environment exists. We can do that through things like rich profiles with pictures, real names, and strong enforcement."[59]

Google and other Internet hosting sites, however, have a possible conflict of interest. When I was being attacked online by my harassers, I begged Google to consider my case. The results? Nothing. Google stonewalled me. It wouldn't remove search results for slanderous sites, blogs, and other malicious content about me. Journalist Jon Ronson's book *So You've Been Publicly Shamed* describes how Google (and presumably other Internet hosting sites) benefit financially from these public blowups, when they cause a spike in search traffic, resulting in increased ad revenue from Google AdWords.[60] "Some platforms want you to publish as much as possible online; a lot don't have incentive to stop the shaming," notes Bradley Shear.[61]

Twitter has perhaps been one of the most criticized platforms

Twitter's Trust and Safety council member and tech journalist Larry Magid says that the difficulty is in balancing the concerns of those being harassed with the freedom to criticize in public arenas. "I think technology can help," says Magid. "I don't think it's entirely responsible for the problem, and I don't think it can solve the problem. I think there are tools that can shield you from things, services to flag certain behaviors. I'm always skeptical of monitoring tools [that] parents feel can protect from cyberbullying. I look at those as potentially helpful, but never a panacea. People are people—you can filter words, but you can't filter thoughts."[68]

At the social news site Reddit, one of the Internet's top fifty most-visited sites, mudslinging had long reigned unchecked by its ragtag assembly of volunteer moderators. In 2015, the site finally introduced an antiharassment policy and reporting system, specifically banning five controversial subreddits, such as "transfags" and "fatpeoplehate," which had more than 150,000 subscribers. (Previously, subreddits such as "jailbait," which posted pictures of underage girls and men beating women, had been removed.) "We will ban subreddits that allow their communities to use the subreddit as a platform to harass individuals when moderators don't take action. We're banning behavior, not ideas," the company, then under CEO Ellen Pao, announced.[69] In April 2016, Reddit implemented a blocking tool that allows users to block out private messages and also gives them the ability to block from their personal view posts in comment threads by users who they deem offensive.[70]

Perhaps what it will take for true change is for tech heads to get a taste of their own medicine. When Pao found herself in the crosshairs of angry Redditors over a firing skirmish, she wrote in her resignation letter that the hateful comments she had received made her "doubt humanity." And Reddit board member Sam Altman chided those users, saying that what they wrote was "sickening." "The reduction in compassion that happens when we're all behind computer screens is not good for the world," he wrote. "People are still people even if there is Internet between you."[71]

Hear! Hear!

CHAPTER 11

A TIME TO HEAL

TEN STEPS TO REBOOTING YOUR LIFE AFTER A DIGITAL SHAMING

"You can insist on a different ending to your story."

—Monica Lewinsky[1]

Do I ever see a time when the Internet will be free and clear of people cyberslamming each other? The realist in me says no. People are human. Some are just plain mean. Sadly, the Internet is full of trolls and vindictive people who actually enjoy social combat, using their keyboards for unsavory purposes. You will always have that woman or man who wants e-venge, cybercriminals waiting to exploit your personal life, or trolls who simply have nothing better to do. As we saw with people minding their own business like C. D. Hermelin (on his typewriter) and ordinary citizens like Sean O'Brien (Dancing Man), and with actresses like

Leslie Jones, or sports reporters like Jen Royle, digital shaming and verbal violence has, for many, even become a blood sport. Online abuse is here to stay.

So, can you ever overcome digital shaming?

Yes.

I've done it. As an adult victim and survivor of cyberharassment, digital shaming, and defamation, I personally understand the deep, dark hole you descend into when you are attacked. I would wake up every morning in a state of panic, afraid of what I might find in my email inbox. The Internet was no longer a safe place—it was a house of horrors. I felt completely alone, fearful, and hopeless. I have frequently said that unless you have been a victim of online abuse, it's almost impossible to relate. But one of the most important things to realize is that *you are not alone*. Digital shaming has happened and there is nothing that can be done to change that. It's time to get your footing and start to reinvent the new you.

As Monica Lewinsky put it in her 2015 TED Talk: "Anyone who is suffering from shame and public humiliation needs to know one thing: you can survive it. I know it's hard. It may not be painless, quick, or easy, but you can insist on a different ending to your story. Have compassion for yourself. We all deserve compassion, and to live both online and off in a more compassionate world."[2] I couldn't agree more. You can *redefine your life* after an online shaming. It happened to you, but it doesn't have to *define* you.

Here's how to move forward.

1. **Get angry.** Permit yourself to be angry; it's okay and it's normal. Use these feelings to energize yourself for what you'll need to do to fight back. It took me nearly a decade to realize that we can't control how people act online, but we can control our own emotions. It's okay to be upset, hurt, and especially mad. Your anger can fuel you forward, whether you decide to fight back, pursue legal recourse, or just pour energy into scrubbing your online persona. For Jen Royle, nasty online comments taught her she was tougher than she even knew. "It made me stronger," she says. "I stick up for myself more now. No one could ever walk on me."[3]

2. **Move forward *with* your shame.** Acceptance is one of the key steps in overcoming digital shame. Life *changes* after a virtual whipping. I don't care what people who haven't walked in these shoes say—we are *not* the same people we once were. It's not that we don't love life or appreciate the things we used to enjoy, but we do have to redefine ourselves. The first step is saying YES. Accept that this has happened to you and that you will have to learn to live with it. Acceptance doesn't mean that you are allowing the cyberslime to win; on the contrary, the digital wisdom and skills you'll develop from this experience will help you repair your virtual landscape and learn that this electronic humiliation won't define you. Digital shaming happened to you, just as it has to many others. It sucks, it hurts, it's an emotional thrashing like no other—but we can survive it and, as cliché as it may sound, *thrive from it.*

"You'll never get justice, you will never get even," says revenge porn victim-turned-advocate Annmarie Chiarini. "Learning how to accept it, that's the key to moving on."[4] And attorney Bradley Shear says, "Sometimes, I have to act like a therapist and tell people, 'It's time to move on, there's nothing more we can do, legally.' You just don't have control [over] how people see you online. [The] best thing is to be the best person you can be."[5] Now let's take those first steps forward.

3. **You matter. So take care of yourself.** I was a take-charge person who would never want to burden anyone else with my problems. Yet when I hit rock bottom, I felt like there was no one I could turn to. After all, who would understand? The humiliation was unfathomable. I was fortunate enough to have one friend who I felt completely secure with; otherwise, I don't know how I would have survived. Having someone you can confide in without judgment is so important when you are dealing with digital shaming. Maybe that means turning to family. Or having a night out with friends to bolster your confidence. But you may also need to seek out professional help. When I was going through this, many therapists were not in tune with how digital shaming was affecting lives. Psychologists are only now realizing that the symptoms of those who have been digitally shamed are very similar to those of PTSD, as Dr. Michele Borba described in chapter 5. Many victims of digital cruelty have shared with me how they have

struggled with depression and anxiety, withdrawing from society, or taking prescription medications. As hard as it is, you have to pick yourself up and get the help you need.

While you're tending to your emotional well-being, distract yourself by getting your body fit. Join a gym, start walking, try water aerobics, treat yourself to a Fitbit. Yes, you can do this. (I didn't say it was easy—it's not.) Get out of your house and move! Unplug and get physical. Chances are very good that the people you are walking with or hiking with don't care about your online persona; they only want to enjoy your company. Or try out items that have a proven calming effect, like adult coloring books, a stress ball, or kinetic sand. "Mine was watching puppy videos, or looking at a picture of my own dog," recalls therapist Samantha Silverberg.[6] Revenge porn activist Holly Jacobs found her salvation in training for a half marathon, listening to music, and reading inspirational quotes. Enjoy your moments.

4. **Get perspective.** Off-line and especially online, there will always be detractors (some will be trolls) who will want to bring you down, particularly if jealousy or other unkind motives are involved. Just because someone hurt you doesn't mean that you are worthless or that you don't have anyone who cares for you. Sean Kosofsky, director of the Tyler Clementi Foundation, insists that you must remember this: "Know that you're loved. You can be experiencing all this intense negativity. People do love you, people are

being cruel because it's just simpler for some to be mean than thoughtful. Sometimes, it's going to get worse before it gets better. Keep some perspective. The kinds of folks attacking you don't matter. They *just don't matter*."[7]

5. **Use laughter.** Nothing seems funny right now, but is there a way to use black humor to see the situation with fresh eyes? Sarah, the woman who was attacked for her comments on *Sex and the Ivy*, says that when her new boyfriend Googled her name and read all the horrible lies her attacker had posted, she had to explain the whole story to him. His conclusion? "If you've pissed off someone that much…maybe you've done something right!"

"Figure out a way to laugh at the situation," agrees Holly Jacobs, who tells of one revenge porn victim who was approached at grocery store by a man inquiring, "Didn't I see you on this site?" Her response? "So… I see *you* watch a lot of porn."[8] The man freaked and skulked off, and she got a good chuckle out of it.

In 2012, when Monica Lewinsky was going through rough times, a good friend sent her some links to bloopers and funny videos to cheer her up. Initially, Monica felt annoyed, but after clicking the email, found herself laughing and watched every single one. "Laughter is sort of this lightning bolt that can help jiggle the system and remind you that as terrible as you're feeling, that you can actually be in a different state," Monica said.[9]

Check out some YouTube videos from the popular,

ongoing *Jimmy Kimmel Live!* segment "Celebrities Read Mean Tweets," where notables from George Clooney to former president Barack Obama drolly read aloud and respond to their worst Twitter abusers. Watching celebrities shrug off their attacks is a good way to learn how the pros deflate the hate.

6. **Find your voice.** When victims email me about what has been therapeutic for them, the most common theme I hear is *writing.* Have you written your story yet? Whether it's on a blog or in your personal journal, expressing your feelings in writing, and sometimes sharing them, even online—the very place where you were harmed—can be cathartic. You'd be surprised how many victims have found a safe space by opening a Blogger account or starting a WordPress blog and telling their story. It may take you time to do this, but when you do, you will find many people waiting to give you support. "Putting it down, typing it down, you take control of your story, and that's really powerful," says Wattpad star Emily Lindin, author of *UnSlut: A Diary and a Memoir.*[10] You don't have to be a professional writer with connections to the publishing world. Even posting your experience on Facebook can be a way to feel that your voice is being heard.

7. **Give back. Pay it forward with small acts of kindness.** You're dealing with your own drama right now, so why am I telling you to start helping others? Because it works! It's that old cliché—when you do good for others, it's usually

the one who is giving that benefits the most. I do say *small acts* of kindness, because you have to budget your time and find your balance without overextending yourself. It can be as simple as sending cyberhugs to people who need them or making time to call a distant relative or friend who you haven't talked to in a long time. Just as keystrokes destroyed you, keystrokes can also lift you up. Social media can transcend grief and sadness. Many organizations offer ways to get involved with promoting kindness in the world. Sign up to be a HeartMobber or pledge to be an Upstander at the Tyler Clementi Foundation, and speak up when you see online bullying.

Off-line community service is just as important and can be key to your recovery. Get involved with organizations you are passionate about! Volunteer, even if it is only an hour or two a week at your favorite charity—you will be amazed at how good you will start feeling. Whether you help underprivileged children, a humane society, or a nursing home, trust me, these people are not concerned about what the Internet is saying about you. Ripple Kindness founder Lisa Currie tells this story: "A tragedy in our family really opened [my eyes] to the power of kindness. As my family was consumed by grief, my husband and I slowly realized that it was the small acts of kindness we were doing for others that were helping us stay connected and making us feel better. The more we did and were involved in helping others, the

more positive our thoughts became, and eventually the
fog started to lift."[11]

8. **Redefine your reputation.** It's time to take the steps
outlined here to rebuild your online presence. Now is
the time to be your most authentic self, so people will
understand that if they see negative content about you,
it is likely just an outlier. Engage your followers with
enthusiasm and positivity. Humanize yourself, making
sure your words are filled with compassion. And finally,
educate others by sharing your wealth of knowledge.
You don't have to be a Harvard graduate with extensive academic credentials; it can be as simple as guiding
someone on how to get wine stains out of a marble
countertop. When it comes to educating others online,
don't hold back—become the person who people in your
field defer to. If you're an interior designer or exceptional
craftsperson, share your designs or tips. You *do* have the
power to change your online narrative. Take heart in
stories of others who have been broken and have found
ways to rebuild their lives again. From Dancing Man to
the Roving Typist, they found ways to turn their negative
online experiences around. You can do this too.

9. **Regain trust.** Many who have been shamed, stalked,
bullied, or harassed online find themselves changed.
You often want to retreat—keep a low profile, limit your
comments, and become very guarded. Basically, your trust
level goes out the window. It took me years to build my

confidence back, open myself up on social media, and let new friendships flourish. It's not an easy task for people who have been slammed online, but it's one you must pursue. Find people with similar passions or interests. I have made wonderful social media friends who have surrounded me with warmth, some who are even more loyal than some real-life friends.

10. **Be shameless.** Are we moving from Shame Nation into a post-shame world—perhaps a *Shameless* Nation? Difficult as it may seem, so many politicians and celebrities seem to bounce back unabashed from a scandal. Plagiarists come out with new books.[12] Girls shrug off the fact that their nude photos are being traded like baseball cards. Can you, too, co-opt this nonchalant attitude? Remember that you are not alone—it's likely that, before long, everyone will suffer their own moment of online mortification. As Emily Lindin noted, with an entire generation growing up sending nudes, how could there be such a thing as a sex scandal? When we all have been the subject of electronic embarrassment, will the shame still sting?[13]

Whether you picked up this book to understand the culture of online hate or to learn how to secure your digital reputation, hopefully you're feeling better equipped and more confident that you can handle any cyberblunders and digital disasters that may come your way. You've learned that digital wisdom is digital survival, to always be selective in your sharing, and that what you post is Public and Permanent. If you do encounter a cyberfoe, you've learned how best to respond, seek out the right resources, and repair your reputation.

What's next?

We have seen the worst we can be within our online communities. Now it is time to turn it around. The choices are many. You can advocate for legal reforms, choose to be an Upstander when you see cruelty, or create a kindness campaign. Together, we can stay skeptical of slurs and instill empathy in our youth. We must take steps to restore civility and turn our Shame Nation into a Sane Nation.

The next time you witness online hate—what will you do?

"Be kind to one another."

—Ellen DeGeneres

ENDNOTES

Introduction

1 Monica Lewinsky, "The Price of Shame," Filmed March 2015, TED video, 22:26, https://www.ted.com/talks/monica_lewinsky_the_price_of_shame?language=en.

2 Maeve Duggan, "Online Harassment," Pew Research Center, October 22, 2014, http://www.pewinternet.org/2014/10/22/online-harassment/.

3 Jake Gammon, "Over a Quarter of Americans Have Made Malicious Online Comments," YouGov, October 20, 2014, https://today.yougov.com/news/2014/10/20/over-quarter-americans-admit-malicious-online-comm/.

4 Jennifer Jacquet, *Is Shame Necessary?: New Uses for an Old Tool* (New York: Pantheon, 2015).

5 Jim Finkle, "Chubb Adds Cyber Bullying Coverage to U.S. Home Insurance Policies," Reuters, March 30, 2016, http://www.reuters.com/article/us-cyber-insurance-bullying-idUSKCN0WW21K.

6 Christie Alderman, interview with author, July 12, 2016.

7 Joseph Diaz and Lauren Effron, "Former CFO on Food Stamps After Controversial Viral Video About Chick-Fil-A," ABC News, March 25, 2015, http://abcnews.go.com/Business/cfo-food-stamps-controversial-viral-video/story?id=29533695.

8 Annie Karni, "Web Site Exposes Bad Tippers in Brooklyn," *New York Post*, May 8, 2011, http://nypost.com/2011/05/08/web-site-exposes-bad-tippers-in-brooklyn/.

9 Clyde Haberman, "Mob Shaming: The Pillory at the Center of the Global Village," *New York Times*, June 19, 2016, http://www.nytimes.com/2016/06/20/us/mob-shaming-the-pillory-at-the-center-of-the-global-village.html.

10   Dan Good, "N.Y. Nurse Who Took Photo of Unconscious Patient's Penis Surrenders
      License," *New York Daily News*, March 29, 2016, http://www.nydailynews.com/new
      -york/nurse-surrenders-license-photographing-patient-penis-article-1.2580259.

11   Lisa Vaas, "'Selfie War' Paramedics Accused of Taking Photos with Unconscious
      Patients," Naked Security (blog), Sophos, July 22, 2016, https://nakedsecurity
      .sophos.com/2016/07/22/selfie-war-paramedics-accused-of-taking-photos-with
      -unconscious-patients/.

12   Dr. Sameer Hinduja, email interview with author, September 21, 2016.

13   Primus Telecommunications Canada Inc., "How Times Have Changed:
      Cyberbullying Outranks Drugs, Teenage Pregnancy and Alcohol as a Top Concern
      of Canadian Parents," news release, January 13, 2015, http://www.marketwired.com
      /press-release/how-times-have-changed-cyberbullying-outranks-drugs-teenage
      -pregnancy-alcohol-as-top-1982373.htm.

14   The National Campaign to Prevent Teen and Unplanned Pregnancy, *Sex and
      Tech: Results from a Survey of Teens and Young Adults*, December 2008, https://the
      nationalcampaign.org/resource/sex-and-tech.

15   Kevin Draper, "NBA Star Draymond Green Sends Out Dick Pic on Snapchat,
      Apologizes," Deadspin, July 31, 2016, http://deadspin.com/nba-star-draymond
      -green-sends-out-dick-pic-on-snapchat-1784613439.

16   Amanda Michelle Steiner, "Iggy Azalea Quits Twitter: 'The Internet Is the Ugliest
      Reflection of Mankind There Is,'" *People*, February 10, 2015, http://people.com
      /celebrity/iggy-azalea-quits-twitter-after-cellulite-criticism-everyone-deserves
      -peace/.

17   Simone Olivero, "Mom Fires Back after Toddler Fat Shamed on Reddit," Yahoo, August
      9, 2016, https://ca.style.yahoo.com/two-people-said-she-unhealthily-000000070.html.

18   Martin Pengelly and Kevin Rawlinson, "Reddit Chief Ellen Pao Resigns after
      Receiving 'Sickening' Abuse from Users," *Guardian*, July 10, 2015, https://www
      .theguardian.com/technology/2015/jul/10/ellen-pao-reddit-interim-ceo-resigns.

19   Larry Magid, interview with author, October 13, 2016.

20   Erin E. Buckels, Paul D. Rapnell, and Delroy L. Paulhaus, "Trolls Just Want to Have
      Fun," *Personality and Individual Differences* 67 (September 2014): 97–102, http://www
      .sciencedirect.com/science/article/pii/S0191886914000324.

21   Lindsay Blackwell, interview with author, October 14, 2016.

22   "Stanford Research Shows that Anyone can Become an Internet Troll," Science Blog,
      last modified February 6, 2017, https://scienceblog.com/491911/stanford-research
      -shows-anyone-can-become-internet-troll/.

23   Dr. Robi Ludwig, interview with author, July 11, 2016.

24   Dr. John Suler, email interview with author, September 26, 2016.

25   Ari Ezra Waldman, "Cybermobs Multiply Online Threats and Their Danger," *New
      York Times*, August 3, 2016, http://www.nytimes.com/roomfordebate/2016/08/03
      /how-to-crack-down-on-social-media-threats/cybermobs-multiply-online-threats
      -and-their.danger.

26   Katja Rost, Lea Stahel, and Bruno S. Frey, "Digital Social Norm Enforcement: Online Firestorms in Social Media," *PLOS ONE* (June 17, 2016), http://dx.doi.org/10.1371 /journal.pone.0155923.

27   Mackenzie Dawson, "How Social Media Is Destroying the Lives of Teen Girls," *New York Post*, February 21, 2016, http://nypost.com/2016/02/21/how-social-media-is -destroying-the-lives-of-teen-girls/.

28   "Why a Dodgy Social Media Profile Is Better Than None," *Undercover Recruiter* (blog), http://theundercoverrecruiter.com/why-no-social-media-presence-is-bigger -red-flag/.

29   CareerBuilder, "Number of Employers Using Social Media to Screen Candidates Has Increased 500 Percent over the Last Decade," news release, April 28, 2016, http://www.careerbuilder.com/share/aboutus/pressreleasesdetail.aspx?ed=12%2F31 %2F2016&id=pr945&sd=4%2F28%2F2016.

30   Representative Katherine Clark, email interview with author, October 23, 2016.

Chapter 1

1   Richard Guerry, interview with author, June 20, 2016.

2   "Passenger Shaming" Instagram page, accessed February 9, 2017, https://www .instagram.com/passengershaming/.

3   *Merriam-Webster Online*, s.v. "shame," accessed February 2, 2017, https://www .merriam-webster.com/dictionary/shame.

4   Devon Kelly, "Bodybuilder Banned from Gym for Mocking Older Woman," Yahoo, December 29, 2016, https://www.yahoo.com/beauty/bodybuilder-banned-from -gym-for-mocking-elderly-woman-203740664.html.

5   David Brooks, "The Shame Culture," *New York Times*, March 15, 2016, http://www .nytimes.com/2016/03/15/opinion/the-shame-culture.html?_r=0.

6   Christine Organ, "Public Shaming Is Out of Control, and It's Hurting All of Us," *Scary Mommy* (blog), http://www.scarymommy.com/public-shaming-hurting-us/.

7   Dr. Michele Borba, interview with author, August 3, 2016.

8   Jillian J. Jordan et al., "Third-Party Punishment as a Costly Signal of Trustworthiness," *Nature* 530 (February 25, 2016): 473–476, http://www.nature.com/nature/journal /v530/n7591/full/nature16981.html?WT.feed_name=subjects_evolutionary-theory.

9   Marie Puente, "Megyn Kelly Tweet-Attacks Shutterfly for 'Lying' about Christmas Card Order," *USA Today*, December 23, 2016, http://www.usatoday.com/story/life /people/2016/12/23/megyn-kelly-tweet-attacks-shutterfly-lying-christmas-card-order /95801328/.

10   Jennifer Jacquet, *Is Shame Necessary?: New Uses for an Old Tool* (New York: Pantheon, 2015), 168–9.

11   Lori Millen, email interview with author, October 11, 2016.

12   Ibid., October 31, 2016.

13   Susan Miller, "Stanford Case Juror to Judge: Shame on You," *USA Today*, June 14, 2016,

http://www.usatoday.com/story/news/nation/2016/06/14/juror-stanford-sentence
-appalled/85855028/.

14   Eve Peyser, "Facebook Apologizes for Taking Down Stanford Rape Case Meme," *Cosmopolitan*, June 7, 2016, http://www.cosmopolitan.com/politics/news/a59565/facebook-apologizes-brock-turner-rape-meme/.

15   Lisa Damour, *Untangled: Guiding Teenage Girls Through the Seven Transitions into Adulthood* (New York: Ballantine, 2016), 141.

16   Meghan Keneally, "Mom's Facebook Shaming Video of 13-Year-Old Goes Viral," ABC News, May 20, 2015, http://abcnews.go.com/US/moms-facebook-shaming-video-13-year-viral/story?id=31190138.

17   Michael Pearson, "Mom Shames Daughter, 13, in Facebook Video," CNN, May 21, 2015, http://www.cnn.com/2015/05/21/living/feat-denver-mom-public-shaming-video/; Deb Stanley and Lindsay Watts, "Denver Mom Val Starks Shames Daughter on Facebook for Posing as 19 and Posting Racy Pictures," Denver 7, May 20, 2015, http://www.thedenverchannel.com/news/local-news/denver-mom-val-starks-shames-daughter-on-facebook-for-posing-as-19-and-posting-racy-pictures05202015.

18   Keneally, "Mom's Facebook Shaming."

19   Diana Kwon, "Put to Shame—and Better for It," *Scientific American*, May 2016, http://www.scientificamerican.com/article/put-to-shame-and-better-for-it/.

20   Damour, *Untangled*, 141.

21   Etan Thomas and Dave Zirin, *The Collision*, radio show audio, May 26, 2016, http://archive.wpfwfm.org/mp3/wpfw_160526_100000collision.mp3.

22   Zoë Corbyn, "Jennifer Jacquet: 'The Power of Shame Is That It Can Be Used by the Weak against the Strong'" *Guardian*, March 6, 2015, https://www.theguardian.com/books/2015/mar/06/is-shame-necessary-review.

23   Jennifer Jacquet, interview with author, November 7, 2016.

24   SammyT1, May 31, 2016, comment on Scott Allen, "'I Didn't Actually Call Her Racist': Ex-NBA Player on Viral Post about Being Denied Seat on Train," *Washington Post*, May 31, 2016, https://www.washingtonpost.com/news/early-lead/wp/2016/05/31/i-didnt-actually-call-her-racist-ex-nba-player-on-viral-post-about-being-denied-seat-on-train/#comments.

25   "Charlotte Driver Endangers 30+ Bicyclists on Ballantyne Commons Pkwy," Reddit, August 25, 2016, https://www.reddit.com/r/Charlotte/comments/4zjzs8/charlotte_driver_endangers_30_bicyclists_on/#bottom-comments.

26   "CBS WKBT News Anchor's On-Air Response to Viewer Calling Her Fat," YouTube, uploaded October 2, 2012, https://www.youtube.com/watch?v=rUOpqd0rQSo.

27   Katie Kindelan, "'Bully' Viewer 'Never Meant to Hurt' Overweight TV Anchor Jennifer Livingston," ABC News, October 5, 2012, http://abcnews.go.com/blogs/entertainment/2012/10/bully-viewer-never-meant-to-hurt-overweight-tv-anchor-jennifer-livingston/.

28   "Furious Mom Has Airport Meltdown after 12-Hour Delay Ruins Disney Cruise," YouTube video, 2:38, posted by *Inside Edition*, April 12, 2016, https://www.youtube.com/watch?v=wjYscFJgl0E.

29   *Good Morning America*'s Facebook page, accessed February 9, 2017, https://www.facebook.com/GoodMorningAmerica/videos/10153549636117061/?hc_ref=SEARCH.

30   Jukin Media, email interview with author, September 26, 2016.

31   Mike Skogmo, "We've Paid More Than $10 Million to Our Video Partners," Jukin Media, December 7, 2016, https://www.jukinmedia.com/blog/view/weve-paid-more-than-million-in-royalties-to-amateur-video-creators.

32   Ryan Bradley, "How Viral Video Companies Can Turn Pizza Rats into Boatloads of Cash," *Guardian*, May 18, 2016, https://www.theguardian.com/media/2016/may/18/pizza-rat-viral-video-jukin-media.

33   Jonathan Skogmo, interview with author, October 6, 2016.

34   Adrian McCoy, "Creators of People of Walmart, Website That Pokes Fun at Walmart Customers, Have Local Ties," *Pittsburgh Post-Gazette*, October 20, 2010, http://www.post-gazette.com/life/lifestyle/2010/10/20/Creators-of-People-of-Walmart-website-that-pokes-fun-at-Walmart-customers-have-local-ties/stories/201010200159.

35   Adam Kipple, Andrew Kipple, and Luke Wherry, interview with author, August 12, 2016.

36   "Michigan Woman Upset Over Photo Posted on 'People of Walmart' Website," Fox News, February 17, 2011, http://www.foxnews.com/tech/2011/02/17/michigan-woman-upset-photo-posted-people-walmart-website.html.

Chapter 2

1   Patrick Ambron, interview with author, June 30, 2016.

2   BrandYourself and Harris Interactive, "Just Google Me: How Our Personal Search Results Affect Our Everyday Relationships, from Who We Do Business With, Who We Vote For, and Even Who We Date," BrandYourself.com, October 15, 2012, http://blog.brandyourself.com/wp-content/uploads/Harris-FULL-Study-Report-3.pdf.

3   Patrick Ambron, "When Parent-Shaming Goes Digital, We're All in the Gorilla Pen," Huffington Post, June 10, 2016, http://www.huffingtonpost.com/patrick-ambron/when-parentshaming-goes-d_b_10357898.html.

4   Patrick Ambron, interview with author, October 20, 2016.

5   Rich Matta, email interview with author, October 13, 2013.

6   Annmarie Chiarini, "I Was a Victim of Revenge Porn. I Don't Want Anyone Else to Face This," *Guardian*, November 19, 2013, https://www.theguardian.com/commentisfree/2013/nov/19/revenge-porn-victim-maryland-law-change.

7   Annmarie Chiarini, interview with author, July 27, 2016.

8   Email interview with Duxbury High School junior, September 15, 2016.

9   Susanna Sheehan, "Police Pursue Photo Suspect," *Duxbury Clipper*, May 11, 2016, https://theduxburyfile.wikispaces.com/file/view/Police%20pursue%20photo%20suspect%205-11-2016.pdf/582859937/Police%20pursue%20photo%20suspect%205-11-2016.pdf.

10   Interview with Duxbury High School mother, August 9, 2016.

11   "Carleton Students Caught with Inappropriate Pictures of Peer," *Winnetka Current*, November 13, 2014, http://www.winnetkacurrent.com/police/carleton-students-caught-inappropriate-pictures-peer.

12   Matthew Clancy, interview with author, September 21, 2016.

13   Jeff R. Temple and HyeJeong Choi, "Longitudinal Associations Between Teen Sexting and Sexual Behavior," *Pediatrics* (October 2014), http://pediatrics.aappublications.org/content/pediatrics/early/2014/09/30/peds.2014-1974.full.pdf.

14   Elizabeth Englander, "Low Risk Associated with Most Teenage Sexting: A Study of 617 18-Year-Olds," *MARC Research Reports* (July 2012), http://vc.bridgew.edu/marc_reports/6/.

15   Englander, "Coerced Sexting and Revenge Porn Among Teens," *Bullying, Teen Aggression, and Social Media* 1, no. 2 (March/April 2015), https://www.researchgate.net/publication/274696549_Coerced_Sexting_and_Revenge_Porn_Among_Teens.

16   Andrew Stephens, interview with author, June 20, 2016.

17   "No Charges Expected in Duxbury High School Nude Photo Scandal," WCVB.com, May 6, 2016, http://www.wcvb.com/article/no-charges-expected-in-duxbury-high-school-nude-photo-scandal/8234630.

18   Liz Brody, "Meet Ashley Reynolds, the Woman Fighting 'Sextortion,'" *Glamour*, July 7, 2015. http://www.glamour.com/story/ashley-reynolds-the-woman-fighting-sextortion.

19   Federal Bureau of Investigation, "Sextortion: Help Us Locate Additional Victims of an Online Predator," FBI.gov, July 7, 2015, https://www.fbi.gov/news/stories/sextortion.

20   Scott Stump, "Newly Crowned Miss Teen USA: I Was a Victim of Cybercrime," *Today*, August 12, 2013, http://on.today.com/1cHHMeF?cid=eml_onsite.

21   Benjamin Wittes et al., "Sextortion: Cybersecurity, Teenagers, and Remote Sexual Assault," Brookings Institution, May 11, 2016, https://www.brookings.edu/research/sextortion-cybersecurity-teenagers-and-remote-sexual-assault/.

22   Ibid.

23   Melissa Pamer, "Massachusetts Man Posed as Justin Bieber to Get Naked Photos from 9-Year-Old LA County Girl, Sheriff's Dept. Says of 'Sextortion' Case," KTLA News, January 16, 2017, http://ktla.com/2017/01/16/massachusetts-man-posed-as-justin-bieber-to-get-naked-photos-from-9-year-old-l-a-county-girl-sheriffs-dept-says-of-sextortion-case/.

24   Tamara Vaifanua, "Utah Mother Shares Experience of Son Who Died by Suicide after 'Sextortion' Scam," Fox 13, January 12, 2017, http://fox13now.com/2017/01/13/utah-mother-shares-experience-of-son-who-committed-suicide-after-sextortion-scam/.

25   Brody, "Meet Ashley Reynolds."

26   Rita Panahi, "Mt Martha Woman Snapped Sunbaking in G-String by Real Estate Drone," *Herald Sun*, November 16, 2014, http://www.heraldsun.com.au/news/victoria/news-story/c3eaaeb6318d7f01dcb4394da968340a.

27   Jim Coyle, "Newfoundland Teen Finds Beautiful Response to Social Media Bullying,"
        Toronto Star, Dec. 3, 2015, https://www.thestar.com/news/canada/2015/12/03
        /newfoundland-teen-finds-beautiful-response-to-social-media-bullying.html.

28   Geoff Bartlett, "'Ugliest Girls' Poll: Student's Response to Cyberbullying Goes Viral,"
        CBC News, Dec. 2, 2015, http://www.cbc.ca/news/canada/newfoundland-labrador
        /lynelle-cantwell-ugliest-girl-facebook-1.3347311.

29   Ibid.

30   Lynelle Cantwell, interview with author, September 11, 2016.

31   "Lynelle Cantwell's Response to 'Ugliest Girls' Poll Earns Torbay Teen Trip to Toronto," CBC
        News, December 7, 2015, http://www.cbc.ca/news/canada/newfoundland-labrador
        /lynelle-cantwell-global-leadership-conference-1.3353835.

32   Cantwell, interview with author, September 20, 2016.

33   MakeupByDreigh's Instagram page, accessed February 13, 2017, https://www
        .instagram.com/makeupbydreigh/?hl=en.

34   Ryan Broderick, "After Discovering She Had Been Turned Into a Cruel Meme, This
        Woman Decided to Speak Out," BuzzFeed, October 8, 2015, https://www.buzzfeed
        .com/ryanhatesthis/this-girl-is-speaking-out-after-her-instagram-photo-became
        -a?utm_term=.hlG7RzWE8#.jlA53Zq2N.

35   Ashley VanPevenage, "My Response to My Viral Meme," YouTube video, 1:50,
        posted by "Ashley VanPevenage," October 6, 2015, https://www.youtube.com
        /watch?v=YqwYTYMtpTg; VanPevenage, interview with author, September 20, 2016.

36   "Let's Give Karen—The Bus Monitor—H Klein a Vacation!," Indiegogo page, accessed
        February 13, 2017, https://www.indiegogo.com/projects/lets-give-karen-the-bus
        -monitor-h-klein-a-vacation—6#/.

37   Charles Ornstein and Jessica Huseman, "Inappropriate Social Media Posts by
        Nursing Home Workers, Detailed," ProPublica, December 21, 2015, https://www
        .propublica.org/article/inappropriate-social-media-posts-by-nursing-home
        -workers-detailed.

38   Charles Ornstein, "Nursing Home Workers Share Explicit Photos of Residents on
        Snapchat," ProPublica, December 21, 2015, https://www.propublica.org/article
        /nursing-home-workers-share-explicit-photos-of-residents-on-snapchat.

39   Galit Breen, interview with author, September 15, 2016.

40   Galit Breen, "12 Secrets Happily Married Women Know," Huffington Post, June
        25, 2014, http://www.huffingtonpost.com/galit-breen/12-secrets-happily-married
        -women-know_b_5528785.html.

41   Galit Breen, interview with author, September 15, 2016.

42   Galit Breen, "IT HAPPENED TO ME: I Wrote an Article about Marriage, and All
        Anyone Noticed Is That I'm Fat," xoJane.com, September 10, 2014, http://www
        .xojane.com/it-happened-to-me/it-happened-to-me-i-wrote-an-article-about
        -marriage-and-all-anyone-noticed-is-that-im-fat.

43   Breen interview.

44   Jaya Saxena, "Woman Known as 'Teacher Bae' Sparks Online Debate about

Appropriate Workwear," The Daily Dot, September 12, 2016, http://www.dailydot.com/irl/teacher-bae-modesty/.

45 Lara Rutherford-Morrison, "The Internet Is Outfit-Shaming a Teacher for Wearing 'Sexy' Clothes & Ignoring That She's Amazing at Her Job," Bustle, September 13, 2016, http://www.bustle.com/articles/183683-the-internet-is-outfit-shaming-a-teacher-for-wearing-sexy-clothes-ignoring-that-shes-amazing-at.

46 Yesha Callahan, "Atlanta Public School System Reprimands #TeacherBae," The Root, September 14, 2016, http://www.theroot.com/blog/the-grapevine/atlanta-public-school-system-reprimands-teacherbae/.

47 Terri Parker, "Mean Moms Bash 'Ugly' Toddlers in Secret Facebook Group," WPBF 25, last modified November 6, 2013, http://www.wpbf.com/article/mean-moms-bash-ugly-toddlers-in-secret-facebook-group/1319522.

48 Talia Koren, "Body-Shaming Has Reached Its Lowest Level after Trolls Called a Baby Fat," Elite Daily, August 12, 2016, http://elitedaily.com/social-news/body-shaming-baby-fat/1580669/.

49 Darren Rovell, "'Stuff Curry' Emerges from Tragedy to Become Internet Sensation," ESPN, May 22, 2016, http://www.espn.com/nba/story/_/id/15644403/stuff-curry-emerges-tragedy-become-internet-sensation.

50 Ibid.

51 Baby Landon Lee's Instagram page, accessed February 21, 2017, https://www.instagram.com/babylandonlee/.

52 Dr. Robi Ludwig, interview with author, July 11, 2016.

53 Patrick Ambron, "When Parent-Shaming Goes Digital, We're All in the Gorilla Pen," Huffington Post, June 10, 2016, http://www.huffingtonpost.com/patrick-ambron/when-parentshaming-goes-d_b_10357898.html.

54 4BoysMother–Melissa Fenton, Writer's Facebook page, accessed February 21, 2017, https://www.facebook.com/4BoysMother/posts/966840756748074.

Chapter 3

1 "#CancelColbert," The Internet Ruined My Life, SyFy Channel, March 9, 2016, http://www.syfy.com/theinternetruinedmylife/videos/1.

2 Doug Oakley, "Newark Teacher Who Wrote Nasty, Threatening Tweets Given Reprimand," Mercury News, August 27, 2014, http://www.mercurynews.com/2014/08/27/newark-teacher-who-wrote-nasty-threatening-tweets-given-reprimand/.

3 J.K. Trotter, "How Twitter Schooled an NYU Professor about Fat-Shaming," The Atlantic, June 3, 2013, http://www.theatlantic.com/national/archive/2013/06/how-twitter-schooled-nyu-professor-about-fat-shaming/313728/.

4 Rachel Brodsky, "Can Anthony Weiner Go to Jail for Sexting a 15-Year-Old Girl?," Rolling Stone, September 22, 2016, http://www.rollingstone.com/culture/news/can-anthony-weiner-go-to-jail-for-sexting-a-15-year-old-w441525.

5 J. D. Miles, "Single Mom Fired Before First Day over Facebook Post," CBS DFW,

April 27, 2015, http://dfw.cbslocal.com/2015/04/27/single-mom-fired-before-first-day-over-facebook-post/.

6   Anna Merlan, "Woman Fired Before First Day of Job for Facebook Post about Hating Job," Jezebel, April 28, 2015, http://jezebel.com/woman-fired-before-first-day-of-job-for-facebook-post-a-1700628295.

7   Sam Biddle, "Justine Sacco Is Good at Her Job, and How I Came to Peace with Her," Gawker, December 20, 2014, http://gawker.com/justine-sacco-is-good-at-her-job-and-how-i-came-to-pea-1653022326.

8   Jon Ronson, "How One Stupid Tweet Blew Up Justine Sacco's Life," New York Times Magazine, February 12, 2015, http://www.nytimes.com/2015/02/15/magazine/how-one-stupid-tweet-ruined-justine-saccos-life.html?_r=0.

9   "Woman Complains That Meal Was 'Ruined' by Heart Attack Victim," Fox News, January 4, 2016. http://www.foxnews.com/leisure/2016/01/04/internet-slams-hairstylist-who-claimed-her-new-years-eve-meal-was-ruined-by/.

10   Shari Rudavsky and Amy Haneline, "Criticism of Kilroy's Illuminates Risks of Facebook Rants," Indianapolis Star, January 4, 2016, http://www.indystar.com/story/life/food/2016/01/04/kilroys-goes-viral-after-response-angry-customer/78252026/?hootPostID=edebf26963214ddde0741f2db6513ce4.

11   Breeanna Hare, "Harvard Professor Sorry for Fighting Restaurant over $4," Eatocracy (blog), CNN, December 11, 2014, http://www.cnn.com/2014/12/10/living/harvard-business-professor-chinese-takeout/.

12   Rachel Zarrell, "What Happens When You Dress as a Boston Marathon Victim and Post It on Twitter," BuzzFeed, November 2, 2013, https://www.buzzfeed.com/rachelzarrell/what-happens-when-you-dress-as-a-boston-marathon-victim?utm_term=.lkYY9ZlB2#.ik068EZQD.

13   Soraya Nadia McDonald, "'Racists Getting Fired' Exposes Weaknesses of Internet Vigilantism, No Matter How Well-Intentioned," Washington Post, December 2, 2014, https://www.washingtonpost.com/news/morning-mix/wp/2014/12/02/racists-getting-fired-exposes-weaknesses-of-internet-vigilantism-no-matter-how-well-intentioned/.

14   Racists Getting Fired's Tumblr page, November 24, 2014, accessed February 19, 2017, http://racistsgettingfired.tumblr.com/post/103693021987/this-post-was-faked-to-frame-brianna-rivera-do.

15   "Recruit Yuri Wright Expelled for Tweets," ESPN, January 20, 2012, http://www.espn.com/college-sports/recruiting/football/story/_/id/7484495/yuri-wright-twitter-posts-cost-college-scholarship.

16   "Yuri Wright," CUBuffs, http://www.wpbf.com/article/mean-moms-bash-ugly-toddlers-in-secret-facebook-group/1319522.

17   "Duke Coach Lays Out Social Media Rules for Players," Third Parent (blog), July 11, 2016, http://thirdparent.com/duke-coach-lays-out-social-media-rules-for-players/.

18   Ben Kercheval, "Texas Tech Coaches Are Using Fake Social Media Accounts to Spy on Players," CBS Sports, August 26, 2016, http://www.cbssports.com/college

-football/news/texas-tech-coaches-are-using-fake-social-media-accounts-to-spy
-on-players/.

19   Madison Park, "Miss Teen USA Questioned about Racial Slur," CNN, August 1, 2016, http://www.cnn.com/2016/08/01/us/miss-teen-usa-slur/index.html?sr=fbCNN 080116miss-teen-usa-slur1006AMStoryLink&linkId=27144372.

20   Dave Quinn, "Miss Teen USA Karlie Hay Says She Is 'Sorry' and 'Ashamed' of N-Word Tweets," People, August 2, 2016, http://people.com/bodies/miss-teen-usa -karlie-hay-sorry-and-ashamed-of-n-word-tweets/.

21   "Oops: Virginia Congressional Candidate Leaves Porn Tabs Open on Facebook Post," Fox News, May 17, 2016, http://www.foxnews.com/politics/2016/05/17/oops -virginia-congressional-candidate-leaves-porn-tabs-open-on-facebook-post.html.

22   Laura Beck, "Report: Cops Finally Find Woman From Dani Mathers's Gym Pic, and She's Allegedly PISSED," Cosmopolitan, September 5, 2016, http://www .cosmopolitan.com/lifestyle/news/a63710/cops-find-woman-dani-mathers-body -shamed-at-gym/.

23   Richard Winton and Veronica Rocha, "Prosecutors Charge Former Playboy Playmate Dani Mathers in Gym 'Body-Shaming' Photo Case," Los Angeles Times, November 4, 2016, http://www.latimes.com/local/lanow/la me body shaming -20161104-story.html.

Chapter 4

1   Megan Sutton, "10 Reasons Renée Zellweger Is Our Hero," Red, September 6, 2016, http:// www.redonline.co.uk/red-women/interviews/10-reasons-renee-zellweger-is-our-hero.

2   "Celebs Who Have Been Body-Shamed," CNN, last modified February 8, 2017, accessed February 14, 2017, http://www.cnn.com/2015/06/05/entertainment/gallery /body-shaming-celebs/.

3   Kirthana Ramisetti, "Candace Cameron Bure Slams Online Trolls on 'The View': 'You Don't Have to Verbally Abuse and Rape Me,'" New York Daily News, October 1, 2015, http://www.nydailynews.com/entertainment/tv/candace-cameron-bure-compares -online-bullying-rape-article-1.2381338.

4   Ben Child, "Carrie Fisher Blasts Star Wars Body Shamers on Twitter," Guardian, December 30, 2015, https://www.theguardian.com/film/2015/dec/30/carrie-fisher -blasts-star-wars-body-shamers-twitter-social-media.

5   Elias Leight, "Paris Jackson Speaks out Against Cyberbullying, Defends Justin Bieber," Rolling Stone, September 15, 2016, http://www.rollingstone.com/music /news/paris-jackson-talks-cyberbullying-defends-justin-bieber-w439989.

6   Jia Wertz, "Grown Adults Are Calling Blue Ivy Ugly on the Internet," Huffington Post, September 14, 2016, http://www.huffingtonpost.com/entry/grown-adults-are -calling-blue-ivy-ugly-on-the-internet_us_57c83f1ee4b06c750dd8e510.

7   Howard Bragman, interview with author, June 28, 2016.

8   Emily Hutchinson, "Carrie Underwood Attacked by Online Bullies, Admits She

Has to 'Have a Barrier Up'" Inquisitr, October 10, 2016, http://www.inquisitr.com/3577394/carrie-underwood-attacked-online-bullies-bullied-barrier-up/.

9 Deborah Evans Price, "Carrie Underwood Says She's 'Incredibly Disappointing to People' When She's Not Performing," Redbook, October 10, 2016, http://www.redbookmag.com/life/interviews/a46273/carrie-underwood-cover-story/.

10 Emy LaCroix, "Wendy Williams: Haters Viciously Mock Her 'Awkward' Childhood Photo," Hollywood Life, December 22, 2016, http://hollywoodlife.com/2016/12/22/wendy-williams-throwback-pic-mocked-haters-tweets/.

11 Rachel McRady, "Lena Dunham Quits Twitter: 'It Wasn't a Safe Space for Me... It Creates Cancerous Stuff Inside You,'" Us Weekly, September 30, 2015, http://www.usmagazine.com/celebrity-news/news/lena-dunham-quits-twitter-it-creates-cancerous-stuff-inside-you-2015309.

12 Jaleesa M. Jones, "Fifth Harmony's Normani Kordei Returns to Twitter to Tackle Cyberbullies," USA Today, September 15, 2016, http://www.usatoday.com/story/life/entertainthis/2016/09/15/fifth-harmonys-normani-kordei-returns-to-twitter-to-tackle-cyber-bullies/90410532/.

13 Tony Hicks, "Justin Bieber Has Returned to Social Media," Mercury News, August 29, 2016, http://www.mercurynews.com/music/ci_30303488/justin-bieber-has-returned-social-media.

14 Samuel Gibbs, "Justin Bieber Quits Instagram after Feud with Selena Gomez," Guardian, August 16, 2016, https://www.theguardian.com/technology/2016/aug/16/justin-bieber-quits-instagram-selena-gomez.

15 Day & Night, "Cyber Bullying Has Turned Jennifer Aniston into a Social Hermit," Daily Express, August 20, 2014, http://www.express.co.uk/celebrity-news/501347/Cyber-bullying-turned-Jennifer-Aniston-into-social-hermit.

16 Aili Nahas, "Giuliana Rancic: I Know I'm Too Thin," People, April 1, 2015, http://www.people.com/article/giuliana-rancic-admits-too-thin.

17 Ree Hines, "Giuliana Rancic on Controversial 'Fashion Police' Comments: It Was Edited Wrong," Today, April 6, 2015, http://www.today.com/popculture/giuliana-rancic-tells-real-story-controversial-comments-t13161.

18 Peter Sblendorio, "Corey Feldman Moved to Tears after 'Today' Show Performance Gets Ridiculed by Viewers," New York Daily News, September 19, 2016, http://www.nydailynews.com/entertainment/music/corey-feldman-cries-today-show-performance-ridiculed-article-1.2798411.

19 Bragman interview.

20 Amanda Hess, "The Latest Celebrity Diet? Cyberbullying," New York Times, October 12, 2016, http://www.nytimes.com/2016/10/13/arts/celebrities-twitter-instagram-cyberbullying-kardashian-swift.html?smid=tw-share&_r=1.

21 Heidi Krupp, interview with author, June 21, 2016.

22 Lisa Respers France, "Renee Zellweger: 'I'm Glad Folks Think I Look Different,'" CNN, October 22, 2014, http://www.cnn.com/2014/10/21/showbiz/celebrity-news-gossip-renee-zellweger-new-look/index.html?hpt=hp_c4.

23   Tanya Ghahremani, "Yes, Renee Zellweger Looks Different, But Please Stop with the Rude Comments," Bustle, October 21, 2014, http://www.bustle.com/articles/45223 -yes-renee-zellweger-looks-different-but-please-stop-with-the-rude-comments.

24   Ree Hines, "Who's the Dad in 'Bridget Jones's Baby'? Renee Zellweger (Still!) Doesn't Know," *Today*, March 25, 2016, http://www.today.com/popculture/who-s-father -bridget-jones-s-baby-even-renee-zellweger-t82526.

25   Owen Gleiberman, "Renee Zellweger: If She No Longer Looks Like Herself, Has She Become a Different Actress?," *Variety*, June 30, 2016, http://variety.com/2016/film /columns/renee-zellweger-bridget-joness-baby-1201806603/.

26   Rose McGowan, "Rose McGowan Pens Response to Critic of Renee Zellweger's Face: 'Vile, Damaging, Stupid, and Cruel,'" *Hollywood Reporter*, July 6, 2016, http:// www.hollywoodreporter.com/news/rose-mcgowan-blasts-varietys-renee-908489.

27   Renée Zellweger, "We Can Do Better," Huffington Post, August 8, 2016, http://www .huffingtonpost.com/renee-zellweger/we-can-do-better_b_11355000.html.

28   Bragman interview.

29   Ty Pendlebury, "Fappening Hacker Gets 18 Months in Prison," CNET, October 28, 2016, https://www.cnet.com/news/fappening-hacker-gets-18-months-in-prison -celebgate/.

30   Edward Lucas, *Cyberphobia: Identity, Trust, Security, and the Internet* (New York: Bloomsbury, 2015), 7.

31   Sam Kashner, "Both Huntress and Prey," *Vanity Fair*, October 20, 2014, http://www .vanityfair.com/hollywood/2014/10/jennifer-lawrence-photo-hacking-privacy.

32   Matthew Rozsa, "Why We Need Milo: Even Offensive Trolls Like Milo Yiannopoulos Are Good for the Left," *Salon*, September 22, 2016, http://www .salon.com/2016/09/22/why-we-need-trolls-even-offensive-clowns-like-milo -yiannopoulos-can-be-good-for-the-left/.

33   Stephanie Petit, "Twitter Bans Racist Trolls Who Attacked Leslie Jones—As CEO Reaches Out to *Ghostbusters* Star," *People*, July 20, 2016, http://people.com/celebrity /leslie-jones-twitter-ceo-jack-dorsey-reaches-out-to-actress/.

34   Joel Stein, "How Trolls Are Ruining the Internet," *Time*, August 18, 2016, http://time .com/magazine/us/4457098/august-29th-2016-vol-188-no-8-u-s/.

35   Kristen V. Brown, "How a Racist, Sexist Hate Mob Forced Leslie Jones off Twitter," Fusion, July 19, 2016, http://fusion.net/story/327103/leslie-jones-twitter-racism/.

36   Nina Golgowski, "Leslie Jones Shoots Down Trolls in Epic 'Saturday Night Live' Segment," Huffington Post, October 23, 2016, http://www.huffingtonpost.com /entry/leslie-jones-trolls-snl_us_580cedb4e4b02444efa3edc4.

37   "Jury in Nude Video Lawsuit Awards Erin Andrews $55 Million," ESPN, March 7, 2016, http://www.espn.com/espn/story/_/id/14923412/jury-awards-erin-andrews-55 -million-lawsuit-stalker-hotels.

38   Ashley Judd, "How Online Abuse of Women Has Spiraled Out of Control," TED Talk, filmed October 2016, https://www.ted.com/talks/ashley_judd_how_online _abuse_of_women_has_spiraled_out_of_control.

39   Ibid.

40   Ashley Judd, "Forget Your Team: Your Online Violence Toward Girls and Women Is
       What Can Kiss My Ass," Mic.com, March 19, 2015, https://mic.com/articles/113226
       /forget-your-team-your-online-violence-toward-girls-and-women-is-what-can
       -kiss-my-ass#.auUbrikrn.

41   Ibid.

42   Judd, "How Online Abuse of Women Has Spiraled."

43   MSNBC, March 16, 2015, "Ashley Judd's Rallying Points," MSNBC video, 6:53,
       posted March 16, 2015, http://www.msnbc.com/thomas-roberts/watch/ashley-judd-s
       -rallying-points-413876803640.

44   Michael Sainato, "Comedian Margaret Cho on Hate-Shaming Internet Trolls,"
       Observer.com, October 21, 2015, http://observer.com/2015/10/comedian-margaret
       -cho-on-hate-shaming-internet-trolls/.

45   Chloe Melas, "Kevin Smith Calls Out Cyberbullies for Attacking His 17-Year-Old
       Daughter," CNN, August 24, 2016, http://www.cnn.com/2016/08/16/entertainment
       /kevin-smith-instagram-daughter-cyberbullying/.

46   Stephanie Petit, "Ariel Winter Slams Body Shamers over Her Graduation Dress: 'I
       Looked HOT in That Dress,'" *People*, June 23, 2016, http://www.people.com/article
       /ariel-winter-graduation-dress-body-shaming-response.

47   Ariel Winter's Twitter page, accessed February 14, 2017, https://twitter.com
       /arielwinter1/status/745784012573073408.

48   Kathryn Lindsay, "Ariel Winter Just Shut Down Her Instagram Shamers with This
       Important Reminder," Hello Giggles, November 14, 2015, http://hellogiggles.com
       /ariel-winter-instagram-shamers/.

49   Jayme Deerwester, "Lady Gaga Sales Spike 1,000% after Super Bowl Halftime
       Show," *USA Today*, last modified February 7, 2016, http://www.usatoday.com
       /story/life/music/2017/02/07/lady-gaga-post-superbowl-half-time-sales-streams
       -spike/97590804/.

50   Lady Gaga, Instagram, https://www.instagram.com/p/BQPMuhPlaBr/.

51   "Lady Gaga Says Harvard Visit More Than Anti-Bullying," ArtisanNewsService,
       March 1, 2012, https://www.youtube.com/watch?v=xVtzQou65aE.

52   Christine Burroni, "Chris Pratt Says Body-Shaming 'Hurts,'" March 23, 2017, http://
       pagesix.com/2017/03/23/chris-pratt-says-body-shaming-hurts/.

53   Lisa Respers France, "'Prison Break' Star Responds to Body-Shaming: 'I Was
       Suicidal,'" CNN, March 29, 2016, http://www.cnn.com/2016/03/29/entertainment
       /wentworth-miller-body-shaming-feat/.

54   The LAD Bible's Facebook page, accessed February 14, 2017, https://www.facebook
       .com/LADbible/posts/2691099820937191.

Chapter 5

1   Lena Chen, "Former Harvard Sex Blogger: My Ex-Boyfriend Leaking Nude Pictures of Me Changed Who I Am—Forever," *Time*, September 3, 2014, http://time.com /3263406/jennifer-lawrence-celebrity-leaked-photos-revenge-porn/.

2   Matt Singer, "Welcome to the 2015 Recruiter Nation, Formerly Known as the Social Recruiting Survey," Jobvite, September 22, 2015, https://www.jobvite.com/blog/welcome -to-the-2015-recruiter-nation-formerly-known-as-the-social-recruiting-survey/.

3   CareerBuilder, "Number of Employers Using Social Media to Screen Candidates Has Increased 500 Percent over the Last Decade," news release, April 28, 2016, http://www.careerbuilder.com/share/aboutus/pressreleasesdetail.aspx?ed=12% 2F31%2F2016&id=pr945&sd=4%2F28%2F2016.

4   "35 Percent of Employers Less Likely to Interview Applicants They Can't Find Online, According to Annual CareerBuilder Social Media Recruitment Survey," press release, CareerBuilder, May 14, 2015, http://www.careerbuilder.com/share/aboutus /pressreleasesdetail.aspx?sd=5%2F14%2F2015&id=pr893&ed=12%2F31%2F2015.

5   "Kaplan Test Prep Survey: College Admissions Officers Say Social Media Increasingly Affects Applicants' Chances," press release, Kaplan Test Prep, February 10, 2017, http://press.kaptest.com/press-releases/kaplan test prep-survey-college-admissions -officers-say-social-media-increasingly-affects-applicants-chances.

6   "Benefits," Varsity Monitor, accessed February 21, 2017, http://varsitymonitor.com /benefits.php.

7   Proskauer, "Proskauer Releases 2014 Social Media in the Workplace Global Study," news release, April 29, 2014, http://www.proskauer.com/news/press-releases /proskauer-releases-2014-social-media-in-the-workplace-global-study/.

8   Kimberly W. O'Connor, Gordon B. Schmidt, and Michelle Drouin, "Helping Workers Understand and Follow Social Media Policies," *Business Horizons* 59, no. 2 (December 2015), https://www.researchgate.net/publication/288889523_Helping _workers_understand_and_follow_social_media_policies.

9   Talia Jane, "An Open Letter to My CEO," Medium.com, February 19, 2016, https:// medium.com/@taliajane/an-open-letter-to-my-ceo-fb73df021e7a#.8o0xzebb7.

10   Jonathan Chew, "Yelp Fired an Employee after She Wrote a Post about Her Lousy Pay," *Fortune*, February 22, 2016, http://fortune.com/2016/02/22/yelp-employee -ceo/.

11   Russell Goldman and Jason Stine, "Star of Domino's Pizza Gross-Out Video Is Sorry," ABC News, May 4, 2009, http://abcnews.go.com/GMA/Business/story?id=7500551.

12   David Chang, "Philly Fast Food Restaurant Employees Talk about Having Sex with Customers, Describe Unsanitary Acts in Viral Video," NBCPhiladelphia .com, October 24, 2016, http://www.nbcphiladelphia.com/news/local/Philadelphia -Checkers-Restaurant-Viral-Video-Sex-Unsanitary-Firing-398240551.html.

13   Kenneth Olmstead, Cliff Lampe, and Nicole B. Ellison, "Social Media and the Workplace," Pew Research Center, June 22, 2016, http://www.pewinternet.org /2016/06/22/social-media-and-the-workplace/.

14   Associated Press, "Mariners Suspend Clevenger Without Pay for Rest of Season," The Big Story, September 23, 2016, http://bigstory.ap.org/article /d50c0bdf6db9428684d37ca43000bd61/mariners-suspend-clevenger-without-pay -rest-season.

15   Johnny Lieu, "Australian Man Gets Himself Fired after Making Sexist Comment on Facebook," Mashable, November 30, 2015, http://mashable.com/2015/11/30/man -fired-sexism-australia/#.zcjs.Jg4Sqm.

16   Megan Levy, "Hotel Worker Michael Nolan Sacked over Facebook Post to Clementine Ford," *Sydney Morning Herald*, December 1, 2015, http://www.smh .com.au/national/hotel-worker-michael-nolan-sacked-over-facebook-post-to -clementine-ford-20151130-glc1y4.html.

17   Ed Mazza, "Georgia Elementary School Educator Fired after Allegedly Calling Michelle Obama a 'Gorilla,'" Huffington Post, October 4, 2016, http://www .huffingtonpost.com/entry/georgia-teacher-fired_us_57f30ff5e4b0d0e1a9a96112.

18   Adam Carlson, "Florida Prosecutor Fired after He Called Orlando a 'Void,' Hours after Pulse Mass Shooting," *People*, June 26, 2016, http://people.com/celebrity /florida-assistant-state-attorney-ken-lewis-fired-for-orlando-facebook-post/.

19   Chris Matyszczyk, "Woman Tags Sister-in-Law on Facebook, Faces Year in Jail," CNET, January 15, 2016, https://www.cnet.com/news/woman-tags-sister-in-law -on-facebook-faces-year-in-jail/.

20   "Your Social Media Could Affect Your Insurance Rates," *InsuranceQuotes* (blog), InsuranceQuotes.org, accessed February 20, 2017, http://www.insurancequotes.org /auto/your-social-media-could-affect-your-insurance-rates/.

21   Kristin Zurek, interview with author, December 21, 2016.

22   "WBI Survey: Workplace Bullying Health Impact," Workplace Bullying Institute, August 9, 2012, http://www.workplacebullying.org/2012-d/.

23   Kate Baggaley, "How Being Bullied Affects Your Adulthood," *Slate*, June 20, 2016, http://www.slate.com/articles/health_and_science/medical_examiner/2016/06/the _lasting_effects_of_childhood_bullying_are_surprisingly_not_all_detrimental.html.

24   Dr. Michele Borba, interview with author, August 3, 2016.

25   Dr. Robi Ludwig, interview with author, July 11, 2016.

26   Chen, "Former Harvard Sex Blogger."

27   Borba interview.

28   Samantha Silverberg, interview with author, July 28, 2016.

29   Michelle Drouin, interview with author, July 19, 2016.

30   Britt McHenry, as told to Abigail Pesta, "ESPN's Britt McHenry: 'I Blame Myself, But the Video Is Not Who I Am,'" *Marie Claire*, December 12, 2016, http://www .marieclaire.com/culture/news/a24045/espn-britt-mchenry-viral-video-apology/.

31   Interview with divorced professor, October 6, 2016.

32   Jen Royle, interview with author, July 20, 2016.

33   David Zurawik, "Q&A: Jennifer Royle Talks about Leaving 105.7 The Fan and Baltimore," Baltimore Sun, December 21, 2011, http://www.baltimoresun.com

/entertainment/tv/z-on-tv-blog/bal-jennifer-royle-talks-leaving-1057-the-fan
-baltimore-20111221-story.html.

34 Royle interview.

Chapter 6

1   Diana Graber, *The Internet Ruined My Life*, SyFy Channel, March 16, 2016.

2   Amanda Lenhart, "Teens, Technology, and Friendships," Pew Research Center, August 6, 2015, http://www.pewinternet.org/2015/08/06/teens-technology-and -friendships/.

3   Stuart Wolpert, "Psychology Study Explains When and Why Bystanders Intervene in Cyberbullying," Medical Press, January 15, 2016, https://medicalxpress.com /news/2016-01-psychology-bystanders-intervene-cyberbullying.html#jCp.

4   Stacey Steinberg, "Sharenting: Children's Privacy in the Age of Social Media," University of Florida Levin College of Law Legal Studies Research Paper Series, Paper No. 16-41 (March 8, 2016), accessed 2/17/17 via SSRN, https://ssrn.com /abstract=2711442.

5   Marie Claire Dorking, "A Teenager Is Suing Her Parents for Posting Embarrassing Childhood Pictures to Facebook," Yahoo, September 15, 2016, https://www.yahoo .com/beauty/a-teenager-is-suing-her-parents-for-posting-embarrassing-childhood -pictures-to-facebook-2–194446501.html?soc_src=social-sh&soc_trk=fb.

6   Allison Cacich, "What Was She Thinking?: Daycare Worker Fired after Posting Crude Photo to Snapchat," *Life & Style*, February 5, 2016, http://www.lifeandstylemag.com /posts/daycare-worker-fired-over-snapchat-post-89970.

7   Brandon Griggs, "5 Types of People You Should Unfriend," CNN, November 17, 2014, http://www.cnn.com/2014/11/17/living/national-unfriend-day/.

8   Jon Ronson, "'Overnight, Everything I Loved Was Gone': The Internet Shaming of Lindsey Stone," *Guardian*, February 21, 2015, https://www.theguardian.com /technology/2015/feb/21/internet-shaming-lindsey-stone-jon-ronson.

9   Becky Bratu, "After Threats, Subway Worker Says Comments on Slain Cops Taken the 'Wrong Way,'" NBC News, May 11, 2015, http://www.nbcnews.com/news/us -news/subway-worker-says-comments-dead-cops-taken-wrong-way-n357251.

10 Tom Sykes, "Rapper G-Eazy Reportedly Caught Snorting Cocaine off Naked Woman's Body," The Daily Beast, August 31, 2016, http://www.thedailybeast.com /articles/2016/08/31/g-eazy-celebrates-britney-spears-vma-show-by-snorting -cocaine-off-woman-s-breasts.html.

11 Al Kamen, "One More Question..." *Washington Post*, December 4, 2008, http:// voices.washingtonpost.com/44/2008/12/one-more-question.html.

12 Rich Matta, email interview with author, October 13, 2016.

13 BrandYourself, "Online Reputation Management: The Ultimate Guide," BrandYourself .com, https://brandyourself.com/online-reputation-management.

14 Matta interview; ReputationDefender, "Reputation Management Strategies: How

to Respond to Negative Comments," *ReputationDefender on Online Reputation Management* (blog), ReputationDefender.com, June 16, 2016, https://www.reputation defenderblog.com/reputation-management-strategies-respond-negative-comments/.

15   Toni Birdsong, "Could Your Social Media History Come Back to Bite You?" *Intel Security* (blog), McAfee.com, August 9, 2016, https://blogs.mcafee.com/consumer /could-your-social-media-history-come-back-to-bite-you/.

16   Alice E. Marwick, Lindsay Blackwell, and Katherine Lo, "Best Practices for Conducting Risky Research and Protecting Yourself from Online Harassment," Data & Society, October 18, 2016, https://datasociety.net/output/best-practices-for -conducting-risky-research/.

17   Theresa Payton, email interview with author, September 29, 2016.

18   Alex Hern, "Mark Zuckerberg Tapes over His Webcam. Should You?," *Guardian*, June 22, 2016, https://www.theguardian.com/technology/2016/jun/22/mark-zuckerberg -tape-webcam-microphone-facebook.

19   Martin Kaste, "Why the FBI Director Puts Tape over His Webcam," NPR, April 8, 2016, http://www.npr.org/sections/thetwo-way/2016/04/08/473548674/why-the-fbi -director-puts-tape-over-his-webcam.

20   Richard Guerry, interview with author, July 20, 2016.

21   Heather Kelly, "What to Do if Your Yahoo Account Was Hacked," CNN, September 22, 2016, http://money.cnn.com/2016/09/22/technology/yahoo-hack-password -tips/.

22   Nominet, "Parents 'Oversharing' Family Photos Online, But Lack Basic Privacy Know-How," news release, September 5, 2016, https://www.nominet.uk/parents -oversharing-family-photos-online-lack-basic-privacy-know/.

23   Janita Docherty, email interview with author, November 16, 2016.

24   Guerry interview.

25   Nancy Jo Sales, "The Marines' 'Slut Pages' Are No Surprise to Your Average High School Student," *Guardian*, March 7, 2017, https://www.theguardian.com /commentisfree/2017/mar/07/marines-slut-pages-no-surprise-average-high-school -student?CMP=share_btn_fb.

26   Brandon T. McDaniel and Michelle Drouin, "Sexting Among Married Couples: Who Is Doing It, and Are They More Satisfied?" *Cyberpsychology, Behavior, and Social Networking* 18, no. 11 (November 2015), http://www.ncbi.nlm.nih.gov /pubmed/26484980.

27   Data & Society, "New Report Shows That 4% of U.S. Internet Users Have Been a Victim of 'Revenge Porn,'" news release, December 13, 2016, https://datasociety.net /blog/2016/12/13/nonconsensual-image-sharing/.

28   Kimberly Truong, "This Is How Often Your Nudes Are Being Shared," Refinery29, February 6, 2017, http://www.refinery29.com/2017/02/139571/nudes-shared-sexting -survey.

29   Christina Gagnier, interview with author, August 19, 2016.

30   Robert S. Weisskirch, Michelle Drouin, and Rakel Delevi, "Relational Anxiety and

Sexting," *The Journal of Sex Research* (May 31, 2016), http://www.tandfonline.com /doi/full/10.1080/00224499.2016.1181147.

31   Michelle Drouin, interview with author, July 19, 2016.

32   Elizabeth Englander, "Coerced Sexting and Revenge Porn Among Teens," *Bullying, Teen Aggression, and Social Media* 1, no. 2 (March/April 2015), https://www.researchgate .net/publication/274696549_Coerced_Sexting_and_Revenge_Porn_Among_Teens.

33   Paul Woolverton, "NC Law: Teens Who Take Nude Selfie Photos Face Adult Sex Charges," *Fayetteville Observer*, September 2, 2015, http://www.fayobserver .com/news/local/nc-law-teens-who-take-nude-selfie-photos-face-adult/article _ce750e51-d9ae-54ac-8141-8bc29571697a.html.

34   Englander, "Coerced Sexting and Revenge Porn."

35   Jared Dublin, "Johnny Manziel's Lawyer Who Sent the Accidental Text Is Off the Case," CBS Sports, June 27, 2016, http://www.cbssports.com/nfl/news/johnny -manziels-lawyer-who-sent-the-accidental-text-is-off-the-case/.

36   Martin County Sheriff's Office's Facebook page, accessed February 16, 2017, https:// www.facebook.com/MartinCountySheriffsOffice/posts/1029984020345474.

37   Lucia Moses, "'We're Looking for Mistresses': How Obit Site Legacy Combats Trolling the Deceased," Digiday, January 18, 2016, http://digiday.com/publishers /looking-mistresses-obit-site-legacy-keeps-trolls-bay/.

38   Justin Ellis, "What Happened after 7 News Sites Got Rid of Reader Comments," Nieman Lab, September 16, 2015, http://www.niemanlab.org/2015/09/what -happened-after-7-news-sites-got-rid-of-reader-comments/.

Chapter 7

1   Beverlee J. McClure, "It's Time to Stand Up to Anonymous Cyber Bullies," *Pueblo Chieftain*, February 25, 2017, http://www.chieftain.com/opinion/ideas/it-s-time -to-stand-up-to-anonymous-cyber-bullies/article_d8fbe3a3-82b2-5051-a4f3 -9e84b0c6b800.html.

2   Lindsay Blackwell, interview with author, October 14, 2016.

3   Mitch Jackson, email interview with author, October 11, 2016.

4   Kate Bigam, email interview with author, January 7, 2017.

5   Melissa Fenton, interview with author, September 7, 2017.

6   Claire Fallon, "Neo-Nazi Trump Supporters Are Going after YA Books Now," Huffington Post, September 23, 2016, http://www.huffingtonpost.com/entry/neo -nazi-trump-supporters-are-going-after-ya-books-now_us_57e42d67e4b08d73 b8303486.

7   *The Internet Ruined My Life*, season 1, episode 1, https://www.youtube.com/watch? v=ohY3Qbi7w0s.

8   Larry Magid, interview with author, October 13, 2016.

9   Jackson interview.

10   Clare O'Connor, "Monica Lewinsky Speaks Out on Ending Online Abuse," Forbes,

October 20, 2014, https://www.forbes.com/sites/clareoconnor/2014/10/20/full
-transcript-monica-lewinsky-speaks-out-on-ending-online-abuse/#7e7cc8b82579.

11   Ibid.

12   Monica Lewinsky, "Meet the New Emoji Tool to Combat Cyberbullying," *Vanity Fair*, February 8, 2016, http://www.vanityfair.com/news/2016/02/monica-lewinsky -emoji-safer-internet.

13   O'Connor, "Monica Lewinsky Speaks Out."

14   Caitlin Seida, "My Embarrassing Picture Went Viral," *Salon*, October 2, 2013, http:// www.salon.com/2013/10/02/my_embarrassing_picture_went_viral/.

15   Ibid.

16   Jessica Valenti, Twitter, July 27, 2016, https://twitter.com/jessicavalenti/status/7583 48786959781888?lang=en%20Jessica%20Valenti%20Twitter%20Feed%20July%20 27,%202016.

17   Catherine Piner, "Feminist Writer Jessica Valenti Takes a Break from Social Media after Threat against Her Daughter," *Slate*, July 28, 2016, http://www.slate.com /blogs/xx_factor/2016/07/28/feminist_writer_jessica_valenti_takes_a_break_from _social_media_after_threat.html.

18   Lindy West, "I've Left Twitter. It Is Unusable for Anyone but Trolls, Robots, and Dictators," *Guardian*, January 3, 2017, https://www.theguardian.com /commentisfree/2017/jan/03/ive-left-twitter-unusable-anyone-but-trolls-robots -dictators-lindy-west.

19   Embry Roberts, "Vlogger Anna Saccone Says Goodbye to YouTube in Powerful Cyberbullying Video," *Today*, July 25, 2016, http://www.today.com/health/vlogger -anna-saccone-says-goodbye-youtube-powerful-cyberbullying-video-t101112.

20   Tom Kludt, "*New York Times* Editor Quits Twitter over Anti-Semitic Tweets," CNN, June 9, 2016, http://money.cnn.com/2016/06/08/media/new-york-times-jon -weisman-twitter/.

21   David French, "Free Speech Is Killing Free Speech," *National Review*, September 15, 2016, http://www.nationalreview.com/article/440073/free-speech-intimidation -tool-internet-trolling-boycotts.

22   Emily May, interview with author, August 1, 2016.

23   Theresa Payton, email interview with author, September 29, 2016.

24   C. D. Hermelin, email interview with author, August 24, 2016.

25   C. D. Hermelin, "I Am an Object of Internet Ridicule, Ask Me Anything," The Awl, September 18, 2013, https://theawl.com/i-am-an-object-of-internet-ridicule-ask -me-anything-1bbb3181da27#.6plw73itt.

26   "I Was a Hated Hipster Meme," video, 4:54, posted under "Bonus Features," February 2014, http://rovingtypistfilm.com.

27   Hermelin email interview.

28   Syfy, *The Internet Ruined My Life*, season 1, episode 6 http://www.syfy.com /theinternetruinedmylife/videos/tweet-at-me-bro.

29   Ibid.

30   Hermelin email interview.

31   Hermelin, "I Am an Object of Internet Ridicule."

32   "Hated Hipster Meme," video.

33   Hermelin email interview.

34   Carol Todd, interview with author, September 3, 2016.

35   Ibid.

36   Lindy West, "Don't Ignore the Trolls. Feed Them until They Explode," Jezebel, July 31, 2013, http://jezebel.com/dont-ignore-the-trolls-feed-them-until-they-explode-977453815.

37   Lindy West, "What Happened When I Confronted My Cruelest Troll," Guardian, February 2, 2015, https://www.theguardian.com/society/2015/feb/02/what-happened-confronted-cruellest-troll-lindy-west.

38   "Mike Gallagher Apologizes for Kelly Clarkson Comments: 'I Am Deeply Ashamed of Myself and I Am Truly, Truly Sorry,'" audio clip, posted by Media Matters, April 6, 2015, http://mediamatters.org/video/2015/04/06/mike-gallagher-apologizes-for-kelly-clarkson-co/203176.

39   Carly Mallenbaum, "Hosts Apologize for 'Fat-Shaming' Kelly Clarkson," USA Today, April 6, 2015, http://www.usatoday.com/story/news/nation-now/2015/04/06/kelly-clarkson-apalogy-chris-wallace/25338779/.

40   Ibid.

41   "Mike Gallagher Apologizes for Kelly Clarkson Comments."

42   "About Mike," The Mike Gallagher Show, accessed February 18, 2017, http://www.mikeonline.com/about-mike/.

43   Corinne Heller, "Kelly Clarkson Mocked about Weight by Fox News' Chris Wallace and Radio Host, Colleague Calls for Apology," E! Online, April 4, 2015, http://www.eonline.com/news/643020/kelly-clarkson-mocked-about-weight-by-fox-news-chris-wallace-and-radio-host-colleague-calls-for-apology.

Chapter 8

1   "About HeartMob," HeartMob, accessed February 18, 2017, https://iheartmob.org/about.

2   Interview with MBA student, October 4, 2016.

3   Lindsay Blackwell, "Trolls, Trouble, and Telling the Difference," Ann Arbor District Library video, 30:04, March 16, 2016, http://www.aadl.org/node/331018.

4   Maeve Duggan, "Online Harassment," Pew Research Center, October 22 2014, http://www.pewinternet.org/2014/10/22/online-harassment/.

5   Eileen French, "Twitter Tackles Harassment with a Better Mute Button," Community Financial News, November 15, 2016, https://www.com-unik.info/2016/11/15/twitter-tackles-harassment-with-a-better-mute-button.html.

6   Mike Bires, email interview with author, February 1, 2017.

7   Theresa Payton, email interview with author, September 29, 2016.

8   Rich Matta, email interview with author, October 13, 2013.

9   Bires email interview.

10   "Online Removal Guide," Cyber Civil Rights Initiative, accessed February 18, 2017, https://www.cybercivilrights.org/online-removal/.

11   "Something Can Be Done! Guide," Without My Consent, accessed February 18, 2017, http://www.withoutmyconsent.org/resources#boxes-box-scbd_welcome_block.

12   Caitlin Dewey, "Why Twitter Gave a Woman's Home Address to Her Cyberstalkers," *Washington Post*, June 8, 2016, https://www.washingtonpost.com/news/the-intersect /wp/2016/06/08/why-twitter-gave-a-womans-home-address-to-her-cyberstalkers/.

13   Mitch Jackson, email interview with author, October 11, 2016.

14   T. C., "What Is the Streisand Effect?," *Economist*, April 16, 2013, http://www .economist.com/blogs/economist-explains/2013/04/economist-explains-what -streisand-effect.

15   Jackson email interview.

16   United States Patent and Trademark Office, "Trademark Application Fee Structure," last modified January 14, 2017, https://www.uspto.gov/trademarks-application -process/filing-online/trademark-application-fee-structure.

17   Bradley Shear, interview with author, July 11, 2016.

18   Rad Campaign, Lincoln Park Strategies, and Craigconnects, "New Poll Details Widespread Harassment Online, Especially on Facebook," news release, accessed February 19, 2017, http://onlineharassmentdata.org/release.html.

19   Jim Edwards, "FBI's 'Gamergate' File Says Prosecutors Declined to Charge Men Believed to Have Sent Death Threats—Even When They Confessed on Video," *Business Insider*, February 16, 2017, http://www.businessinsider.com/gamergate-fbi -file-2017-2/#here-the-fbi-notes-that-gamergate-activists-are-organising-on-a -platform-that-is-also-used-to-distribute-child-pornography-20.

20   Crash Override, "So You've Been Doxed: A Guide on What to Do Next," http://www .crashoverridenetwork.com/soyouvebeendoxed.html.

21   U.S. Congressman Katherine Clark, "Clark Bill Aims to Combat Dangerous 'Swatting' Hoaxes," news release, November 18, 2015, http://katherineclark.house .gov/index.cfm/press-releases?ID=F71DAD9F-18E6-4B66-8B11-384911DE591B.

22   Hannah Levintova, "This Congresswoman Has Plans to Stop Online Harassment," *Mother Jones*, September/October 2016, http://www.motherjones.com/politics/2016 /09/katherine-clark-fight-against-internet-trolls-gamergate.

23   Katherine Clark, interview with author, October 25, 2016.

24   Christina Gagnier, interview with author, August 19, 2016.

25   Mike Bires, email interview with author, September 25, 2016.

26   Graeme Wood, "Scrubbed," *New York* Magazine, June 16, 2013, http://nymag.com /news/features/online-reputation-management-2013-6/.

27   Rich Matta, email interview with author, October 13, 2016.

28   Jessica Bennett, "One Family's Fight Against Grisly Web Photos," *Newsweek*, April 24, 2009, http://www.newsweek.com/one-familys-fight-against-grisly-web-photos-77275.

29   Greg Hardesty, "Family Gets $2.4 Million over Grisly Crash Images," *Orange County Register*, January 30, 2012, http://www.ocregister.com/articles/family-337967-catsouras -nikki.html.

30   Bennett, "One Family's Fight."

31   Patrick Ambron, interview with author, June 30, 2016.

32   BrandYourself, "Online Reputation Management: The Ultimate Guide," BrandYourself .com, https://brandyourself.com/online-reputation-management.

33   Theresa Avila, "Leslie Jones Rallies Support for Gabby Douglas after Online Bullying," *New York* Magazine, August 16, 2016, http://nymag.com/thecut/2016/08 /leslie-jones-rallies-support-for-gabby-douglas.html.

34   Kate Bigam, email interview with author, January 7, 2017.

35   Emily May, interview with author, August 1, 2016.

36   "About the Project," TrollBusters, accessed February 18, 2017, http://www.troll-busters .com/.

37   Nancy Davis Kho, "A Conversation with TrollBusters Founder, Dr. Michelle Ferrier," *EContent*, June 10, 2016, http://www.econtentmag.com/Articles/News/News-Feature /A-Conversation-with-TrollBusters-Founder-Dr-Michelle-Ferrier-111320.htm.

38   Ernie Smith, "Anti-Troll Group Wants to Study Journalists' Tweets," *Associations Now*, September 19, 2016, http://associationsnow.com/2016/09/trollbusters-wants -study-journalists-tweets/.

39   Dan Raisbeck, email interview with author, September 27, 2016.

40   "About Us," Crisis Text Line, accessed February 18, 2017, www.crisistextline.org.

41   Sara Ashley O'Brien, "Facebook Wants to Get Smarter about Suicide Prevention," CNN Tech, March 1, 2017, http://money.cnn.com/2017/03/01/technology/facebook -suicide-prevention/.

42   Zachary Jason, "Game of Fear," *Boston* Magazine, May 2015, http://www.boston magazine.com/news/article/2015/04/28/gamergate/3/.

Chapter 9

1   Cate Matthews, "Taylor Swift Takes Time out of Media Whirlwind to Send Heartfelt Message to Bullied Teen," Huffington Post, September 3, 2014, http://www .huffingtonpost.com/2014/09/03/taylor-swift-bullied-fan_n_5760148.html.

2   Holly Jacobs, interview with author, August 9, 2016.

3   Ibid.

4   Emily Lindin, "Discover Your Confidence—Emily Lindin at TEDxYouth@Toronto," YouTube video, 14:27, posted by TEDxYouth, February 17, 2014, https://www.youtube .com/watch?v=36l-UwlO9D4.

5   Emily Lindin, interview with author, July 1, 2016.

6   "Shared Stories," The UnSlut Project, accessed February 19, 2017, http://www .unslutproject.com/shared-stories.html.

7   Monica Lewinsky, "Bullied in Brooklyn: How We Failed Daniel Fitzpatrick,"

Vanity Fair, August 18, 2016, http://www.vanityfair.com/culture/2016/08/daniel
-fitzpatrick-school-bullies.

8   Sean Kosofsky, interview with author, August 18, 2016.

9   "Upstander Pledge," Tyler Clementi Foundation, accessed February 19, 2017, http://
tylerclementi.org/pledge/.

10   Justin W. Patchin, "Summary of Our Cyberbullying Research (2004–2016),"
Cyberbullying Research Center, November 26, 2016, http://cyberbullying.org
/summary-of-our-cyberbullying-research.

11   Kosofsky interview.

12   Alex Napoliello, "Dharun Ravi Pleads Guilty to Attempted Invasion of Privacy in
Tyler Clementi Case," NJ.com, October 27, 2016, http://www.nj.com/middlesex
/index.ssf/2016/10/dharun_ravi_court_appearance.html.

13   Carol Todd, interview with author, September 3, 2016.

14   Amanda Todd, "My Story: Struggling, Bullying, Suicide, Self Harm," YouTube video,
8:54, posted September 7, 2012, https://www.youtube.com/watch?v=vOHXGNx
-E7E.

15   Diane Strandberg, "Extradition to Proceed in Amanda Todd Case," Tri-City News,
October 2, 2015, http://www.tricitynews.com/news/extradition-to-proceed-in-amanda
-todd-case-1.2096989.

16   Todd interview.

17   Kate Wheeling, "The Dancing Man Is an Example of Both Fat Shaming and
the Internet's Glory," Pacific Standard, March 6, 2015, https://psmag.com/the
-dancing-man-is-an-example-of-both-fat-shaming-and-the-internet-s-glory
-141b2904496f#.97d0dgtlw.

18   Nadia Khomami, "Fat-Shamed 'Dancing Man' Gets Own Back at Star-Studded
Hollywood Party," Guardian, May 25, 2015, https://www.theguardian.com
/technology/2015/may/25/dancing-man-fat-shamed-cyberbullies-hollywood-party.

19   Sean O'Brien, video interview with author, July 26, 2016.

20   Hope Leigh, email interview with author, August 8, 2016.

Chapter 10

1   David Allison, "Diana: The Legacy," Huffington Post, last modified February 2,
2016, www.huffingtonpost.com/david-allison/diana-the-legacy_b_1844945.html.

2   Diane Swanbrow, "Empathy: College Students Don't Have as Much as They Used
To," Michigan News, May 27, 2010, http://ns.umich.edu/new/releases/7724-empathy
-college-students-don-t-have-as-much-as-they-used-to; Sara H. Konrath, Edward
H. O'Brien, and Courtney Hsing, "Changes in Dispositional Empathy in American
College Students Over Time: A Meta-Analysis," Personality and Social Psychology
Review 15, no. 2 (June 24, 2016): 180–198.

3   Michele Borba, interview with author, August 3, 2016.

4   Joann S. Lublin, "Companies Try a New Strategy: Empathy Training," Wall Street

Journal, June 21, 2016, http://www.wsj.com/articles/companies-try-a-new-strategy -empathy-1466501403.

5   Michele Borba, *Unselfie: Why Empathetic Kids Succeed in Our All-About-Me World* (New York: Touchstone, 2016), 58–60.

6   Galit Breen, "Raising a Digital Kid without Having Been One," TEDx video, 13:09, posted December 28, 2015, https://www.youtube.com/watch?v=aBbzlN64ib0&t=20s.

7   "Reach of Leading Social Media and Networking Sites Used by Teenagers and Young Adults in the United States as of February 2017," Statista: The Statistics Portal, February 2017, https://www.statista.com/statistics/199242/social-media-and-networking -sites-used-by-us-teenagers/.

8   Mike Bires, email interview with author, September 25, 2016.

9   Ibid.

10   Lisa Currie, email interview with author, October 11, 2016.

11   Matt Soeth, interview with author, July 27, 2016.

12   Molly Thompson and Lauren Paul, interview with author, September 27, 2016.

13   "Kind Campaign Info, Programs, and Stats," Kind Campaign, accessed February 19, 2017, https://www.kindcampaign.com/wp-content/uploads/2016/05/kind-campaign -info-programs-and-stats.pdf.

14   Greenwich Compliments's Facebook page, accessed February 19, 2017, https://www .facebook.com/groups/1728914447322667/?pnref=story.

15   Email interview with Greenwich Compliments founder, October 19, 2016.

16   Email interview with Supportive Guy, August 28, 2016.

17   Lauren Levy, "This Mom Taught Her Daughter the Ultimate Lesson about Bullying— With a Tube of Toothpaste," PopSugar, August 21, 2016, http://www.popsugar.com /moms/Toothpaste-Bullying-Lesson-42236017?utm_campaign=share&utm_medium =d&utm_source=moms.

18   Kristen Layne's GoFundMe page, accessed February 20, 2017, https://www .gofundme.com/k0x9n8.

19   Christine Coppa, "Teen Bullied over Her Prom Dress Has the Last Laugh," Yahoo, March 26, 2015, https://www.yahoo.com/news/teen-bullied-over-her-prom-dress -has-the-last-114594522852.html?ref=gs.

20   Jennifer O'Neill, "Parents Fight Bullying App with Messages of Love," Yahoo, March 25, 2015, https://www.yahoo.com/news/parents-fight-bullying-app-with-messages -of-love-114592572012.html.

21   Sameer Hinduja, interview with author, July 29, 2016.

22   "Video: We Are All Daniel Cui," Facebook Stories video, 3:15, posted by Peter Jordan, accessed February 19, 2017, https://www.facebookstories.com/stories/1921 /video-we-are-all-daniel-cu.

23   Dan Misener, "Everyone Lies on the Internet, According to New Research," CBC News, August 24, 2016, http://www.cbc.ca/news/technology/misenere-internet-lies -1.3732328.

24   Christina Gagnier, interview with author, August 19, 2016.

25   Jacqueline Cain, "The Trolls Descend on Centre Street Café," *Boston* Magazine, August 3, 2016, http://www.bostonmagazine.com/restaurants/blog/2016/08/03/centre-stree-cafe-review-trolls/.

26   Jake Gammon, "Americans Rely on Online Reviews Despite Not Trusting Them," YouGov, November 24, 2014, http://today.yougov.com/news/2014/11/24/americans-rely-online-reviews-despite-not-trusting/.

27   Vince Sollitto, "Protecting Free Speech: Why Yelp Is Marking Businesses That Sue Their Customers," *Yelp Official Blog* (blog), July 25, 2016, https://www.yelpblog.com/2016/07/protecting-free-speech-yelp-marking-businesses-sue-customers.

28   Bradley Shear, interview with author, July 11, 2016.

29   Howard Rheingold, "Crap Detection Mini-Course," YouTube video, 24:33, posted February 20, 2013, http://rheingold.com/2013/crap-detection-mini-course/.

30   Larry Magid, interview with author, October 13, 2016.

31   Soeth interview.

32   "Stop Cyberbullying Day Annual Report 2015," The Cybersmile Foundation, accessed February 20, 2017, https://www.cybersmile.org/wp-content/uploads/Stop-Cyberbullying-Day-2015-Annual-Report.pdf.

33   "Assessing Cyber Civics," *Cyber Civics Blog*, August 26, 2016, http://www.cybercivics.com/blog.

34   Diana Graber, interview with author, September 13, 2016.

35   Soeth interview.

36   "European Privacy Requests for Search Removals," Google, accessed February 19, 2017, https://www.google.com/transparencyreport/removals/europeprivacy/.

37   Amar Toor, "Facebook, Twitter, Google, and Microsoft Agree to EU Hate Speech Rules," The Verge, May 31, 2016, http://www.theverge.com/2016/5/31/11817540/facebook-twitter-google-microsoft-hate-speech-europe.

38   Theresa Payton, email interview with author, September 29, 2016.

39   Magid interview.

40   Danielle Keats Citron, *Hate Crimes in Cyberspace* (Massachusetts: Harvard University Press, 2014).

41   Shear interview.

42   Jackie Speier, interview with author, October 18, 2016.

43   "Revenge Porn: More Than 200 Prosecuted under New Law," BBC, September 6, 2016, http://www.bbc.com/news/uk-37278264.

44   Representative Katherine Clark, email interview with author, October 23, 2016.

45   Ann Friedman, "Katherine Clark Is Taking on the Trolls," *Elle*, July 13, 2016, http://www.elle.com/culture/tech/a37728/katherine-clark-harassment-abuse-legislation/.

46   Levintova, "This Congresswoman Has Plans."

47   Clark email interview.

48   Eric Chiu, "Twitter Trolls, Online Abuse: Ex-CEO Says Company Failed to Stop User Harassment," International Business Times, February 1, 2017, http://www.ibtimes.com/twitter-trolls-online-abuse-ex-ceo-says-company-failed-stop-user-harassment-2484930.

49 "Civil Comments," Civil, accessed February 20, 2017, https://www.getcivil.com /comments/.

50 Andy Greenberg, "Inside Google's Internet Justice League and Its AI-Powered War on Trolls," *Wired*, September 19, 2016, https://www.wired.com/2016/09/inside-googles -internet-justice-league-ai-powered-war-trolls/.

51 Mary Aiken, *The Cyber Effect* (New York: Spiegel & Grau, 2016), 131.

52 "Amanda Project: Gamified Anti-Bullying App Wins First Place at Microsoft Imagine Cup," August 8, 2016, http://domaingang.com/domain-news/amanda-project-gamified -anti-bullying-app-wins-first-place-at-microsoft-imagine-cup/.

53 "What Is ReThink?," ReThink, accessed February 20, 2017, http://www.rethinkwords .com/whatisrethink.

54 Elyse Wanshel, "Teen Makes 'Sit With Us' App That Helps Students Find Lunch Buddies," Huffington Post, September 12, 2016, http://www.huffingtonpost.com /entry/teen-creates-app-sit-with-us-open-welcoming-tables-lunch-bullying_us _57c5802ee4b09cd22d926463.

55 "Block Together," Block Together, accessed February 20, 2017, https://blocktogether .org/.

56 Lindsay Blackwell, interview with author, October 14, 2016.

57 Nancy Jo Sales, *American Girls: Social Media and the Secret Lives of Teenagers* (New York: Vintage, 2016), 135.

58 Graber interview.

59 Ari Ezra Waldman, email interview with author, September 26, 2016.

60 Jon Ronson, *So You've been Publicly Shamed* (New York: Riverhead Books, 2015).

61 Shear interview.

62 Charlie Warzel, "'It Only Adds to the Humiliation'—How Twitter Responds to Harassers," Buzzfeed, September 22, 2016, https://www.buzzfeed.com/charliewarzel /after-reporting-abuse-many-twitter-users-hear-silence-or-wor?utm_term= .tkVJozvOm#.php2GMAPr.

63 Payton interview.

64 "Progress on Addressing Online Abuse," *The Official Twitter Blog* (blog), Twitter, November 15, 2016, https://blog.twitter.com/2016/progress-on-addressing-online -abuse.

65 John Kennedy, "Twitter Creates Its Version of a Mute Button in Latest Anti-Abuse Move," Silicon Republic, March 1, 2017, https://www.siliconrepublic.com/life/twitter -mute-feature.

66 Elana Premack Sandler, "Instagram Takes on Suicide Prevention," *Psychology Today*, October 26, 2016, https://www.psychologytoday.com/blog/promoting-hope -preventing-suicide/201610/instagram-takes-suicide-prevention.

67 Associated Press, "Twitter: We've Shut Down More Than 360,000 Accounts Since 2015 for Promoting Terror, Extremism," CBS New York, August 18, 2016, http:// newyork.cbslocal.com/2016/08/18/twitter-terror-accounts-shut-down/.

68 Magid interview.

69　　Seth Fiegerman, "Reddit Bans Toxic Communities in Biggest-Ever Crackdown on Hate Speech," Mashable, June 10, 2015, http://mashable.com/2015/06/10/reddit -bans-5-subreddits/#yrhDay2GKGqm.

70　　Christina Warren, "Reddit Launches New Block Tools to Help Temper Harassment," Mashable, April 6, 2016, http://mashable.com/2016/04/06/reddit-blocking-tools /#GCfo6GTyRuqK.

71　　Martin Pengelly and Kevin Rawlinson, "Reddit Chief Ellen Pao Resigns after Receiving 'Sickening' Abuse from Users," Guardian, July 10, 2015, https://www .theguardian.com/technology/2015/jul/10/ellen-pao-reddit-interim-ceo-resigns.

Chapter 11

1　　Monica Lewinsky, "The Price of Shame," Filmed March 2015, TED video, 22:26, https://www.ted.com/talks/monica_lewinsky_the_price_of_shame?language=en.

2　　Ibid.

3　　Jen Royle, interview with author, July 20, 2016.

4　　Annmarie Chiarini, interview with author, July 27, 2016.

5　　Bradley Shear, interview with author, July 11, 2016.

6　　Samantha Silverberg, interview with author, July 28, 2016.

7　　Sean Kosofsky, interview with author, August 18, 2016.

8　　Holly Jacobs, interview with author, August 9, 2016.

9　　Bystander Revolution, "Monica Lewinsky's Challenge: Make 'Em Laugh," September 30, 2015, https://www.youtube.com/watch?v=gCJJHAw6s_E.

10　　Emily Lindin, interview with author, July 1, 2016.

11　　Lisa Currie, email interview with author, October 11, 2016.

12　　Jennifer Senior, "Review: Jonah Lehrer's 'A Book about Love' Is Another Unoriginal Sin," New York Times, July 6, 2016, http://www.nytimes.com/2016/07/07/books /jonah-lehrer-a-book-about-love-review.html?_r=0.

13　　Lindin interview.

RESOURCES

Amanda Todd Legacy Society—www.amandatoddlegacy.org
Born This Way Foundation—www.bornthisway.foundation
BrandYourself—www.brandyourself.com
Bystander Revolution—www.bystanderrevolution.org
CiviliNation—www.civilination.org
ConnectSafely—www.connectsafely.org
Crash Override Network—www.crashoverridenetwork.com
Crisis Text Line—www.crisistextline.org
#CureTheHate—www.curethehate.com
CyberCivics—www.cybercivics.com
Cyber Civil Rights Initiative—www.cybercivilrights.org
Cyberbullying Research Center—cyberbullying.org
Cybersmile Foundation—www.cybersmile.org
Cyberwise—www.cyberwise.org
Diana Award Anti-Bullying Campaign—www.antibullyingpro.com/support-centre
DMCA Defender—www.dmcadefender.com
Fortalice Solutions—www.fortalicesolutions.com
Hack Harassment—www.hackharassment.com
HeartMob—www.iheartmob.org
#ICANHELP—www.icanhelpdeletenegativity.org
Institute for Responsible Online and Cell Phone Communication—www.iroc2.org

Kind Campaign—www.kindcampaign.com
National Center for Missing & Exploited Children—www.missingkids.com
Online SOS—www.onlinesosnetwork.org
Peyton Heart Project—www.thepeytonheartproject.org
Recl@im the Internet—www.reclaimtheinternet.com
Reputation—www.reputation.com
ReputationDefender—www.reputationdefender.com
Ripple Kindness Project—www.ripplekindness.org
Social Assurity—www.socialassurity.com
Spark Kindness—www.sparkkindness.org
Stay Safe Online—www.staysafeonline.org
STOMP Out Bullying—www.stompoutbullying.org
Streaming.Lawyer—www.streaminglawyer.wordpress.com
TrollBusters—www.troll-busters.com
Tyler Clementi Foundation—www.tylerclementi.org
Undox.me—www.undox.me
UnSlut Project—www.unslutproject.com/
Without My Consent—www.withoutmyconsent.org
Women's Speech Project—www.wmcspeechproject.com

INDEX

ACKNOWLEDGMENTS

FROM SUE SCHEFF

It takes a team to create monumental change, and when it comes to curbing cruelty, especially online, I could not have asked for better team players.

First and foremost, I must thank my literary agents—Joelle Delbourgo for believing in this project, and Jacquie Flynn for taking it to the perfect home. My editor at Sourcebooks, Grace Menary-Winefield, is amazing; she dedicated herself to bringing *Shame Nation* to its fullest potential with her unwavering commitment and razor-sharp instincts. My sincerest gratitude also goes to Dominique Raccah, founder and CEO of Sourcebooks, who loved this book from the very beginning and championed it with the same passion and care that took Sourcebooks from a fledgling enterprise to one of the most respected and innovative publishing houses in the industry. May the gods of publishing shine as brightly on all of you as you have on me!

An extra-special thanks goes to Monica Lewinsky, whose TED Talk, "The Price of Shame," motivated me over the long and challenging road it takes to finish any work of substance. I'm confident Monica's TED Talk, with more than ten million views—which I encourage everyone to watch (www.ted.com/talks/monica_lewinsky_the_price_of_shame)—has changed the lives of countless others who have been inflicted with the pain of digital cruelty. I'm honored to share this stand against cyberbullying and shaming with Monica, and her support of *Shame Nation* is appreciated tremendously.

This book is filled with remarkable experts and contributors. Most are colleagues, and some I'm privileged to call friends: Theresa Payton, Howard Bragman, Richard Guerry, Sameer Hinduja, Bradley Shear, Mitch Jackson, Robi Ludwig, Larry Magid, Janita Docherty, Diana Graber, Galit Breen, Annmarie Chiarini, Matt Soeth, Carol Todd, Emily Lindin, Michelle Drouin, Mike Bires, Alan Katzman. With more than twenty-five experts and contributors, words cannot express how extremely grateful I am to each and every one for their support, their generosity of time, and their invaluable expertise. They are collectively committed to making a tamer and kinder cyberspace, and much is owed to them for what progress has been made, and will continue to be made, with their help.

This book would never have been written without my talented and respected cowriter, Melissa Schorr. Melissa spent endless hours, weeks, and months going over and above as a tireless researcher, interviewer, and wordsmith. What good fortune to have her on this team. Melissa, take a bow!

Through my odyssey of hard-earned digital wisdom, I've had the privilege of gaining so much knowledge from the tech industry's best. Not only from the experts in this book, but also from others who I would like to personally thank here, since they have influenced and empowered me over the years: Mary Kay Hoal, Chris Duque, Toni Birdsong, Tito De Morais, Justin Patchin, Linda Criddle, Andrea Weckerle, Katie Greer, Michael Fertik, and Joe Yeager, and my apologies to anyone I've missed. I thank you all.

It's those special people in our lives, the quiet cheerleaders who are supportive even when we're distracted (or stick around while you disappear within the pages of a manuscript), who make us truly rich. Love you always Donna Rotunno, Carol R., Joan D., Cari D., Peggy Gachet, Isabelle Zehnder, Ross Ellis, Christine McGlade, Cynthia Lieberman, and Elaine Griffin (my forever plane buddy). And what would I do without my walking buddy and good friend, Laurie, for pulling me away from the screen to get some much-needed fresh air and sand under my toes. I also want to acknowledge a Facebook group of supporters who probably don't realize how much they impact my mornings as they cheerfully include me in their "beachie" hellos and wish everyone a wonderful day. Virtually. Yes, social media and the Internet have many positive influences on life and can reveal the best of human behavior too. And so, here's a shout-out to Maddy, Lisa W., Kenny,

Phyll, Vicki, Jeri, Deborah, Debi, Paiger, Jean, Lorraine, Rossella, Lisa M., Mary, Doreen, Ginny, Gail, Sharon, Evelyn, B. J., Ashlee, Cheryl, Gladys, Marilyn, Christy, Kendra, Sarah, Harry… And again, my apologies if I missed anyone in this wonderful group.

There is one extra-special person who was particularly instrumental in making this book happen. Michele Borba is not only someone who writes about compassion and empathy; she embodies it through her friendships. Michele has been there for me for years, through laughter (and tears), and when my days were darkest, she had my back. Everyone needs a champion like Michele Borba in their corner. I call her my entourage… I love you to pieces… #MBWY. You are the best!

Finally, to all those who have been shamed or are struggling with digital abuse or online hate, I dedicate this book to you. After I wrote *Google Bomb*, I heard your voices. Please know that you are *not* alone. My greatest hope is that *Shame Nation* will provide doors for you to open and options for you to turn to that weren't around even a few years ago.

Time may be a great healer, but being shamed often clings to us for the rest of our lives—if we let it. Bullies, cyber or otherwise, are mean and they are cowards who lack character. Eleanor Roosevelt got it right when she said, "Only a man's character is the real criterion of worth."

ACKNOWLEDGMENTS
FROM MELISSA SCHORR

It is a rare opportunity to work on a book that is timely, fascinating, and potentially impactful on the lives of so many, and for this, I am forever grateful to my agent and champion, Jacqueline Flynn, for bringing me this project, as well as for her continual support and keen editorial insights. Thanks to our fabulous editor at Sourcebooks, Grace Menary-Winefield, and her team, who saw the vision in this project and made the manuscript shine and the cover sing. Eternal thanks to my dedicated coauthor, Sue Scheff, for trusting me to translate her passion and wisdom onto the page. Working with her was a pleasure and an honor! This book wouldn't have been possible without the generosity of those who shared their expertise, the advocates for the hard work they do every day, and the victims, who bravely shared their stories for the greater good.

On a personal note, I'd like to thank my friends and my writing network, with a special shout-out to Monica Tesler and Michelle Deninger, who served as both sounding boards and hiking buddies. None of this would be possible without the support of my loving family: my mother, Thelma Grossman; the entire Cohen crew; and my husband, Gary, for granting me the gift of time. Finally, to my daughters, Alexa and Charlotte: I hope this makes your world—online and off—just a little freer of hate and shame.

ABOUT THE AUTHORS

Sue Scheff is a nationally recognized author, speaker, parent advocate, and Internet safety expert. In 2001, Sue founded Parents' Universal Resource Experts, Inc., an informational resource that has helped thousands of parents struggling with challenging teenagers. Her books *Wit's End! Advice and Resources for Saving Your Out-of-Control Teen* (2008) and *Google Bomb: How the $11.3M Verdict Changed the Way We Use the Internet* (2009)—coauthored by the late John Dozier, one of the country's leading Internet attorneys—offer a wide range of prescriptive advice for parents and teens in both the real world and cyberspace.

Since she began her work as an advocate, Sue's name and voice have become synonymous with helping others who have been victims of online abuse as well as educating people of all ages about the importance of responsible digital citizenship and protecting their online reputations. As a leader in the movement against cyberbullying, she focuses on teaching others how to avoid virtual cruelty and how to effectively react when it occurs.

In addition to her efforts on the national and international level, Sue believes in the importance of remaining active in her local community; she has held various community service positions and is a member of her local rotary club.

Today she is a sought-after expert who has been featured on major media outlets including *CBS This Morning*, ABC News, *20/20*, *The Rachael Ray Show*, *Dr. Phil*, CNN, *Anderson Cooper 360*, *CBS Nightly News*, Lifetime, Fox News, CBC, BBC, *Dr. Drew HLN*, *CNN Headline News*, and *InSession Court TV*, and noted in major publications such as *USA Today*, *AARP*, *Parenting Magazine*, the *Wall Street Journal*, *Washington Post*, *Miami Herald*, *Forbes*, *Sun-Sentinel*, *Asian Tribune*, *San Francisco Chronicle*, *LA Times*, and many more.

You can follow Sue on Twitter @SueScheff and Facebook.com /sue.scheff.author. To learn more about Sue and the *Shame Nation* book, visit suescheff.com and shamenationbook.com.

Melissa Schorr is a widely published freelance journalist and the author of two young adult novels: *Identity Crisis* and *Goy Crazy*. Currently a contributing editor at the *Boston Globe Sunday Magazine*, Schorr has also served as a stringer for *People* magazine, a columnist for the *Las Vegas Sun*, and a health writer for ABC News. She attended Northwestern University's Medill School of Journalism and was awarded a Knight Science Journalism

Fellowship at MIT. She currently lives outside Boston with her husband, their two daughters, and their dog, Bailey. Find her on the web at www.melissaschorr.com.